Lecture Notes in Artificial Intelligence   9165

Subseries of Lecture Notes in Computer Science

LNAI Series Editors

Randy Goebel
University of Alberta, Edmonton, Canada
Yuzuru Tanaka
Hokkaido University, Sapporo, Japan
Wolfgang Wahlster
DFKI and Saarland University, Saarbrücken, Germany

LNAI Founding Series Editor

Joerg Siekmann
DFKI and Saarland University, Saarbrücken, Germany

More information about this series at http://www.springer.com/series/1244

Petra Perner (Ed.)

Advances in Data Mining

Applications and Theoretical Aspects

15th Industrial Conference, ICDM 2015
Hamburg, Germany, July 11–24, 2015
Proceedings

 Springer

Editor
Petra Perner
IBaI
Leipzig
Germany

ISSN 0302-9743 ISSN 1611-3349 (electronic)
Lecture Notes in Artificial Intelligence
ISBN 978-3-319-20909-8 ISBN 978-3-319-20910-4 (eBook)
DOI 10.1007/978-3-319-20910-4

Library of Congress Control Number: 2015942611

LNCS Sublibrary: SL7 – Artificial Intelligence

Springer Cham Heidelberg New York Dordrecht London

Springer International Publishing AG Switzerland is part of Springer Science+Business Media
(www.springer.com)

Preface

The fifteenth event of the Industrial Conference on Data Mining ICDM was held in Hamburg (www.data-mining-forum.de) running under the umbrella of the World Congress on "The Frontiers in Intelligent Data and Signal Analysis, DSA 2015" (www.worldcongressdsa.com).

After the peer-review process, we accepted 16 high-quality papers for oral presentation. The topics range from theoretical aspects of data mining to applications of data mining, such as in multimedia data, in marketing, in medicine and agriculture, and in process control, industry, and society. Extended versions of selected papers will appear in the international journal *Transactions on Machine Learning and Data Mining* (www.ibai-publishing.org/journal/mldm).

In all, ten papers were selected for poster presentations and six for industry paper presentations, which are published in the ICDM Poster and Industry Proceeding by ibai-publishing (www.ibai-publishing.org).

In conjunction with ICDM, three workshops were run focusing on special hot application-oriented topics in data mining: the Workshop on Case-Based Reasoning (CBR-MD), Data Mining in Marketing (DMM), and I-Business to Manufacturing and Life Sciences (B2ML). All workshop papers are published in the workshop proceedings by ibai-publishing house (www.ibai-publishing.org).

A tutorial on Data Mining, a tutorial on Case-Based Reasoning, a tutorial on Intelligent Image Interpretation and Computer Vision in Medicine, Biotechnology, Chemistry and Food Industry, and a tutorial on Standardization in Immunofluorescence were held before the conference.

We were pleased to give out the best paper award for ICDM for the seventh time this year. There are three announcements mentioned at www.data-mining-forum.de. The final decision was made by the Best Paper Award Committee based on the presentation by the authors and the discussion with the auditorium. The ceremony took place during the conference. This prize is sponsored by ibai solutions (www.ibai-solutions.de), one of the leading companies in data mining for marketing, Web mining, and e-commerce.

We would like to thank all reviewers for their highly professional work and their effort in reviewing the papers.

We also thank the members of the Institute of Applied Computer Sciences, Leipzig, Germany (www.ibai-institut.de), who handled the conference as secretariat. We appreciate the help and understanding of the editorial staff at Springer Verlag, and in particular Alfred Hofmann, who supported the publication of these proceedings in the LNAI series.

Last, but not least, we wish to thank all the speakers and participants who contributed to the success of the conference. We hope to see you in 2016 in New York at the next World Congress on "The Frontiers in Intelligent Data and Signal Analysis, DSA 2016" (www.worldcongressdsa.com), which combines under its roof the following three events: International Conferences Machine Learning and Data Mining, MLDM, the Industrial Conference on Data Mining, ICDM, and the International Conference on Mass Data Analysis of Signals and Images in Medicine, Biotechnology, Chemistry and Food Industry, MDA.

July 2015   Petra Perner

Organization

Chair

Petra Perner — IBaI Leipzig, Germany

Program Committee

Ajith Abraham	Machine Intelligence Research Labs, USA
Andrea Ahlemeyer-Stubbe	ENBIS, The Netherlands
Brigitte Bartsch-Spörl	BSR Consulting GmbH, Germany
Orlando Belo	University of Minho, Portugal
Shirley Coleman	University of Newcastle, UK
Jeroen de Bruin	Medical University of Vienna, Austria
Antonio Dourado	University of Coimbra, Portugal
Geert Gins	KU Leuven, Belgien
Warwick Graco	ATO, Australia
Aleksandra Gruca	Silesian University of Technology, Poland
Pedro Isaias	Universidade Aberta, Portugal
Piotr Jedrzejowicz	Gdynia Maritime University, Poland
Martti Juhola	University of Tampere, Finland
Janusz Kacprzyk	Polish Academy of Sciences, Poland
Mehmed Kantardzic	University of Louisville, USA
Mineichi Kudo	Hokkaido University, Japan
Dunja Mladenic	Jozef Stefan Institute, Slovenia
Eduardo F. Morales	INAOE, Ciencias Computacionales, Mexico
Armand Prieditris	Newstar Labs, USA
Rainer Schmidt	University of Rostock, Germany
Victor Sheng	University of Central Arkansas, USA
Kaoru Shimada	Section of Medical Statistics, Fukuoka Dental College, Japan
Gero Szepannek	Santander Consumer Bank, Germany

Contents

X   Contents

Business Intelligence and Customer Relationship Management

Improving the Predictive Power of Business Performance Measurement Systems by Constructed Data Quality Features? Five Cases

Markus Vattulainen[(✉)]

University of Tampere, Tampere, Finland
markus.vattulainen@gmail.com

Abstract. Predictive power is an important objective for current business performance measurement systems and it is based on metrics design, collection and preprocessing of data and predictive modeling. A promising but less studied preprocessing activity is to construct additional features that can be interpreted to express the quality of data and thus provide predictive models not only data points but also their quality characteristics. The research problem addressed in this study is: can we improve the predictive power of business performance measurement systems by constructing additional data quality features? Unsupervised, supervised and domain knowledge approaches were used to operationalize eight features based on elementary data quality dimensions. In the case studies five corporate datasets Toyota Material Handling Finland, Innolink group, 3StepIt, Papua Merchandising and Lempesti constructed data quality features performed better than minimally processed data sets in 29/38 and equally in 9/38 tests. Comparison to a competing method of preprocessing combinations with the first two datasets showed that constructed features had slightly lower prediction performance, but they were clearly better in execution time and easiness of use. Additionally, constructed data quality features helped to visually explore high dimensional data quality patterns. Further research is needed to expand the range of constructed features and to map the findings systematically to data quality concepts and practices.

Keywords: Business performance measurement system · Data quality · Preprocessing · Feature construction · Predictive classification

1 Introduction

Current business performance measurement systems such as balanced scorecard aim to advance from the measurement of past results to prediction of future ones [21]. Achieving the goal is not an easy task as critics have pointed out [33].

Enabling and restricting prediction performance are the base processes of metrics design, data collection and manipulation, and data analysis [13]. Standard data manipulation (i.e. preprocessing) objective is to identify and transform problematic data quality issues such as missing values. Instead of transformations, an alternative method

© Springer International Publishing Switzerland 2015
P. Perner (Ed.): ICDM 2015, LNAI 9165, pp. 3–16, 2015.
DOI: 10.1007/978-3-319-20910-4_1

to improve prediction performance would be to incorporate data point quality characteristics as additional features the rationale being that predictive algorithms would learn not only from data points but also from the quality characteristics of those points.

Incorporation would lessen the need of preprocessing, which is estimated to be the most time-consuming phase in analysis [35], error-prone task even for the expert and may require computationally expensive preprocessing combinations [9, 10]. Thus, the practical utility of constructed data quality features in the context of predictive modeling of business performance measurement system data would be to increase effectiveness as accuracy of predictions, efficiency as time needed to preprocess the data and easiness of use as less demanding expertize requirements. Also, constructed data quality features could improve the analyst's understanding of the interrelatedness of data quality problems in the data.

In knowledge discovery research feature construction is related to feature selection and feature extraction, of which the latter two are used to reduce data dimensionality. The literature review showed approaches to feature construction generally [15] but not how to construct features with the aim to express the quality characteristics of data points specifically in the business performance measurement system data context. A significant gap in current research on constructed data quality features is uncertainty, whether constructed data quality features can improve predictive classification accuracy. Positive outcome would imply dependency between data production process and the target feature to be predicted.

The research problem addressed in this paper is: can we improve the predictive power of business performance measurement system data by adding constructed data quality features to them? The research problem contains three sub problems: 1. What are the elementary constructed data quality features? 2. Are constructed features important in the prediction task? 3. Do constructed data quality features increase prediction accuracy?

In operationalizing the research question several choices were made. First, fitness for purpose (i.e. performance in prediction task) is taken as definition of data quality and issues regarding the concept of data quality (see [37]) are not further elaborated. The constructed features and their mappings have arisen from the modelling practice in the cases and are not as such based on any theory of data quality. Secondly, the research question is limited to actual business performance measurement system data as they are and exclude any external information about data points or data collection process thus allowing analysis in a situation in which external metadata is not available. Lastly, predictive power is operationalized as predictive classification performance consisting of accuracy of predictions (bias). This operationalization excludes other types of prediction such as time series forecasting or discrete time models. Focus on predictive classification is motivated by less demanding model assumption testing requirements.

Eight constructed data quality features were identified and operationalized including common data quality problems such as missing values, outliers and low within data point variance. In the five cases studied constructed data quality features raised to the top 10 features according to random forest variable importance [4] and performed better in 29/39 and equally on 9/38 tests in classification accuracy using several classifiers. The results suggest that adding constructed data quality features to business performance measurement system data can improve predictive power. Additionally, the cases demonstrated that constructed data quality features can reveal significant non-random and high-dimensional data quality patterns in data.

The rest of the paper is structured as follows. Section 2 presents the related research. Data quality is an extensively studied issue in databases and statistics, but no research was found on constructing data quality features for business performance measurement systems data or how these perform in predictive classification task. In Sect. 3 the research method used is described. In Sect. 4 constructed data quality features for business performance measurement systems data are conceptually constructed based on the cases studied and existing ideas found in literature review and then operationalized. In Sect. 5 five business cases are used to evaluate the performance of constructed data quality features in predictive classification task. High dimensional data quality patterns are also visually explored for three of the cases. Section 6 summarizes the findings and their implications. Further research needs are identified.

2   Related Research

The systematic literature review was conducted following Kitchenham et al. [23] guidelines. The key words were derived from the research question above: preprocessing, data quality, feature construction, predictive classification and business performance measurement system. Data bases used were CiteSeerX, IEEE Explore, Springer Link, Wiley Online Library, EBSCOhost, Science Direct and Google Scholar. Source quality criteria included peer-reviewed articles, journal citation index and timeliness (published after 2000). Outside of key words significant articles were tracked backwards by their references.

Of the competing base processes to improve predictive power, namely metrics design, collection and manipulation of data and predictive modelling, the preprocessing process is the least studied and thus expected to yield novel insight and advances. Also, several papers identifying future priorities for knowledge discovery [11, 24, 42, 44] identify data preprocessing among the most important research questions.

Preprocessing is recognized to consist of data cleaning, data integration, data reduction and data transformation [16] and includes several approaches explored by Pyle [35], Kochanski et al. [27], Abdul-Rahmana et al. [1] and Kotsiantis et al. [29]. Specifically, there are techniques for outlier detection (see. [7, 18, 25]), missing values [19], class imbalance [31] and feature selection [15, 28].

Guyon and Eliseeff [15] note that constructed features can be used to insert application domain specific knowledge into data analysis in contrast to automated construction of features through adding, subtraction, multiplication, division etc. Construction can be supported by using neural network hidden layers, clustering and matrix factorization.

In the literature studied three generic approaches can be discerned, of which the first one concerns single data problems and competing algorithmic methods to solve them. This approach is characterized by reliance on exploratory data analysis conducted by data analysis expert, who works by identifying one data quality problem at a time and solving it before moving to next one (Fig. 1).

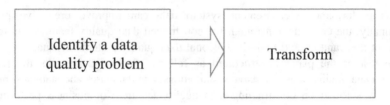

Fig. 1. Single data problem identification and transform

A common problem of missing values can demonstrate the first approach. Identification of missing values is a non-trivial task [3]. The origins of missing values are in the data production process, which is often opaque to the analyst. At the process output level, missing values can be represented in various ways. These include explicit "NA" or implicit "-" and "(blank)" to name a few. Misleading missing value representations can include "0" or categorical "999". After identification a choice is made whether to proceed in data analysis with missing values. Some learning algorithms such as decisions trees can handle missing value but many can not. Also, the data analyst should find out whether missing values are missing at random or not. This can be done for example by labeling each data point as missing or not and then running hierarchical clustering, but often it requires acquisition of domain knowledge. If missing values are missing at random and data set size allows, missing values rows can be discarded. Alternatively, missing values can be imputed with several methods starting from simple imputation of the mean value of the feature for each missing value to complex modeling of missing values.

The second approach to preprocessing is based on preprocessing combinations and acknowledges dependencies between data problem corrections and complexity that arises from that. Thus, the emphasis changes from the exploratory data analysis and the expert analyst to more computational and automated approach. Here several data quality problems are identified at a time and their transformations are assumed to affect one other (Fig. 2).

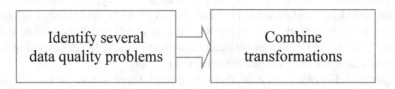

Fig. 2. Preprocessing combinations

Crone et al. [9] and Engel et al. [10] studied preprocessing combinations, interaction effects between single data problem corrections and the latter the difficulty of using exploratory data analysis in determining what kind of preprocessing should be done. Combinations studies and industry standards such as CRISP-DM [8] suggest an order for preprocessing, but permuting the order may prove to be useful [39].

The third approach and the gap in the current research for business performance measurement system data preprocessing can seen as absense of:

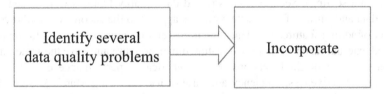

Fig. 3. Incorporation

In the example case of missing values this would mean counting the number of missing values in a data point and then testing in the prediction task, whether that count as a constructed feature can improve prediction performance. Adding other constructed features like outliers would enable the visualization of dependencies between data quality problems, which can not easily be done with preprocessing combinations.

3 Research Method

The research method can be seen to consist of the steps taken to solve the question [20]. In the current study the steps taken included 1. identification of data quality dimensions, 2. operationalization of data quality dimensions, 3. computation of feature importance, 4. computation of classification performance, 5. visual exploration of data quality patterns and 6. comparison to competing methods.

Based on the modelling done in the five cases and literature review data quality dimensions were iteratively identified.

In the operationalization of the dimensions unsupervised, supervised and domain knowledge based approaches were used. Elementary operationalization were preferred over sophisticated ones. The operationalization itself consisted of two steps: operationalized definitions for the selected dimensions and the actual computation of constructed features from data. The operationalized definitions and their mapping to dimensions varied from the evident such as missing value share as incompleteness of data to interpretative such as neighborhood diversity as representativeness. Actual computation of features as well as feature importance and classification performance were done by using the R package Caret [26].

Feature importance was computed with random forest algorithm [4], which evaluates features in the context of the other features and appeared in the analysis of the cases to be more stable than competing stochastic gradient boosting [14] and faster than Relief [22]. Technically, feature importance was computed as: "For each tree, the prediction accuracy on the out-of-bag portion of the data is recorded. Then the same is done after permuting each predictor variable. The difference between the two accuracies are then averaged over all trees, and normalized by the standard error." (Random Forest R Package manual).

For the computation of classification performance three versions of each dataset were created. Firstly, the original data was used as a benchmark and only minimally required preprocessing was done (i.e. imputation of missing values with class mean, centering and scaling). Secondly, constructed data quality features were added to the original data and thirdly, feature selection was applied to the second version to control the effect of adding features [2]. Thus, the third versions aim to act as controllers of an unwanted external factor of increased feature dimensionality and have the same number of features as the original versions. More generically, performance evaluation was divided into effectiveness, efficiency and ease of use following March & Smith [32].

To test the impact of constructed data quality features both base classifies and ensemble classifiers [6] were used. Base classifiers included linear discriminant analysis (abbreviated as LDA), logistic regression (MNOM), k-nearest neighbors (KNN), CART decision tree (TREE) [36] and support vector machine with radial basis kernel (SVM) (see [17]). The ensemble classifiers were random forest bagging (RF) [4], stochastic gradient boosting (GBM) [14] and simple majority vote (VOTE) [30]. Stacking [43] was left out for reasons specified below.

Classification accuracy was used as a performance measurement construct over kappa, since there was no significant class imbalance problem. For validation, holdout sampling was done with 70 % training set share and repeated 1008 times. Holdout for was preferred over cross-validation due to small sample sizes and over bootstrap sampling with replacement due to simplicity.

The constructed features as such without adding them to the original data formed data set for visualization of data quality patterns. There are numerous methods for reducing the dimensionality of data quality features for visualization purposes [34]. For the case demonstrations hierarchical clustering of data quality features by correlation as distance measure was done.

Lastly, with two datasets comparison was done against the competing method of preprocessing combinations.

4 Eight Constructed Data Quality Features

Selected concepts and their operationalization were as follows (Table 1): *Missing-ValueShare.* For each row the number of missing values was counted and their sum was divided by the total number of features. Missing values are a major issue in business data sets [11] and map to the data quality dimension of completeness (see [41]). *Distance-ToNearest.* The data was min-max normalized and Euclidean distance to the nearest data point was computed. This operationalized the data quality concept of redundancy (e.g. duplicates), which is related to the entity identification problem [45]. *LengthOfIQR.* The data was min-max normalized and length of interquartile ranges of the resulting row vectors were computed as opposed to the conventional use of IQR for features. This maps to accuracy and what is called flatness of reply in the survey methodology. *UnivariateOutlier.* For each data point the number of features that were univariate outliers (+1,5 IQR definition) was summed and divided by the total number of features. *MultivariateOutlier.* The multivariate outliers scores were computed by using Torgo's algorithm [38]. Outliers were mapped to data quality concept of accuracy.

Table 1. Constructed data quality features

Feature name	Operationalization	Data quality mapping
MissingValueShare	Number of features with missing values divided by total number of features	Completeness
DistanceToNearest	Euclidean distance to the nearest data point	Redundancy (duplicates)
LenghtOfIQR	Length of interquartile range for a data point (see below).	Accuracy
UnivariateOutlierShare	Number of features with univariate outliers divided by total number of features	Accuracy
MultivariateOutlierScore	Multivariate outlier score as defined by Torgo (2010)	Accuracy
NearestPointPurity	Nearest data point is of same class or not	Representativeness
NeighborhoodDiversity	The ratio of the most frequent class to the second frequent in the neighborhood	Representativeness
LinearDependency	Data point specific residuals from linear model representing the assumed relationship between selected two features	Consistency
RandomValues	RandomValues from range [0,1]	–

NearestPointPurity. Binary value, whether the nearest point is of same class or not. *NeighborhoodDiversity.* The ratio between the counts of classes within the neighborhood of 1/n radius. The nearest point and neighborhood map to data representativeness within a class. LinearDependency. A data point specific residual after setting a linear regression line to data between two variables that are assumed to be linearly dependent in domain knowledge. *LinearDependency* maps to consistency and introduces business rules to control data quality. *RandomValues.* For testing the sanity of the findings, a feature consisting of random values [0,1] was generated.

There were also identified but unoperationalized feature candidates such as *UniqueValueShare*, the number of features of all features that a specific data point has unique value, and *DistanceToClusterCenter*, the Euclidean distance of a data point to cluster center and *AcceptedRange,* binary whether a value is within accepted range as defined in domain specific business rules. There were also alternatives to operationalize a particular dimension such as using LOF [5] to calculate multivariate outlier scores.

5 Demonstration: Cases

To demonstrate constructed data quality features in business performance measurement system context five real corporate datasets were gathered. The main selection criteria were the availability of the financial target feature in the data and at least two of the three other dimensions (customer, process and HR) included. The collection of datasets has unique value in a sense that it includes core business metrics such as sales and profitability measures broken down to personal, customer account and process run levels (Table 2).

Table 2. The cases

Company	Industry	Target feature	Unit of observation	Sample size	Number of features	Length of time-series
Toyota material handling Finland	Industrial services	Personal sales volume	Technicians	48	23	2 years monthly
Innolink group	Sales services for businesses	Customer account growth	Customers	304	14	2 years
3StepIt	IT leasing	Customer account margin	Customers	344	47	5 years
Papua merchandising	Sales services for retail	Process margin	Process runs	253	10	1 year, daily
Lempesti	Cleaning	Process turnover	Process runs	253	8	1 year, daily

5.1 Hierarchical Clustering of Constructed Features

For three cases hierarchical clustering of constructed features was done by using hclust function from ClustOfVar R package (Figs. 4, 5 and 6).

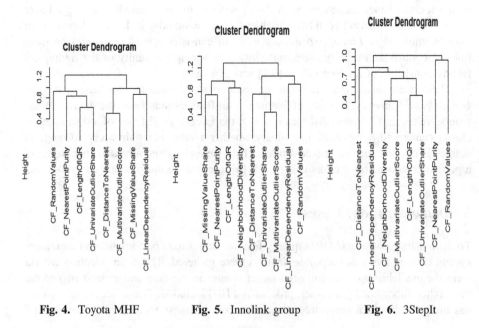

Fig. 4. Toyota MHF   **Fig. 5.** Innolink group   **Fig. 6.** 3StepIt

5.2   Random Forest Variable Importance

Variable importance was computed by using randomForest and varImp functions from the R package Randomforest. The most important features for each case are presented below. Prefix CF denotes that a feature is a constructed data quality feature. (see Tables 3, 4, 5, 6 and 7).

Table 3. 3StepIt

Feature	Importance
rentalmargin	100
COMPANY.INFOCHANGES. IN.ASSET	92.85
AVG_LENGTH.lease.contract. months.	76.94
Personnel.Count.Mean	56.8
Quick.Ratio.Mean	45.63
COMPANYINFO.CHANGES.FOR DEVICES.IN.ASSET	43.02
totalsalescount	42.51
totalendedcount	35.29
CF_NeighborhoodDiversity	32.12
Turnover.Mean	31.56

Table 4. Innolink group

Feature	Importance
OperatingProfit2012	100
CF_LinearDependencyResidual	92.78
Turnover2011	90.08
QuickRatio2011	83.49
OperatingProfit2011	81.99
Turnover2012	78.01
EquityRatio2012	76.21
EquityRatio2011	73.9
QuickRatio2012	69.02
CF_LengthOfIQR	67.29

Table 5. Papua merchandising

Feature	Importance
Tasklength	100
Tasksperresource	14.19
Taskvolume	9.02
CF_NeighborhoodDiversity	7.23
CF_LinearDependencyResidual	6.85
CF_RandomValues	4.4
CF_MultivariateOutlierScore	2.5
CF_DistanceToNearest	2.46
CF_LengthOfIQR	2
Customervolume	1.79

Table 6. Lempesti

Feature	Importance
Tasklength	100
CF_LinearDependencyResidual	32.05
Tasksperresource	31.92
Taskvolume	29.97
CF_NeighborhoodDiversity	17.34
Customervolume	17.07
Minutevolume	14.93
CF_LengthOfIQR	9.09
CF_DistanceToNearest	7.1
CF_RandomValues	5.89

5.3   Misclassification Error

Misclassification error was computed by using ce function from the R package Metrics and standard deviation by R base function sd. The left columns in the table are results for original, only minimally processed datasets. (see Tables 8, 9, 10, 11 and 12) The right columns are the mixed (original + constructed datasets) datasets with the number of features reduced to the same as in the original datasets. The middle columns are the mixed datasets.

Table 7. Toyota material handling Finland

Feature	Importance	Feature	Importance
FINANCIAL.Q1Sales	100	HR.TrainingTime	12,9
HISTORY.PartsSales	58.1	PROCESS.AvailabilityRatio	12.7
HISTORY.TotalSales	42.74	CF_LinearDependencyResidual	12.66
HISTORY.ServiceSales	21.47	FINANCIAL.PartsToServiceRatio	11.64
FINANCIAL.Budget	18.23	HR.ExtraHours	10.74
CF_NearestPointPurity	16,5		
CF_NeighborhoodDiversity	7.23	Taskvolume	29.97
CF_LinearDependencyResidual	6.85	CF_NeighborhoodDiversity	17.34
CF_RandomValues	4.4	Customervolume	17.07
CF_MultivariateOutlierScore	2.5	Minutevolume	14.93
CF_DistanceToNearest	2.46	CF_LengthOfIQR	9.09
CF_LengthOfIQR	2	CF_DistanceToNearest	7.1
Customervolume	1.79	CF_RandomValues	5.89

Table 8. 3StepIt

	Original	Features added	Added and selected
LDA	0.33 ± 0.04	0.35 ± 0.04	0.32 ± 0.04
KNN	0.4 ± 0.05	0.39 ± 0.04	0.36 ± 0.04
TREE	0.34 ± 0.05	0.34 ± 0.05	0.34 ± 0.05
SVM	0.35 ± 0.03	0.35 ± 0.04	0.32 ± 0.04
MNOM	0.1 ± 0.03	0.13 ± 0.04	0.09 ± 0.03
RF	0.27 ± 0.04	0.27 ± 0.04	0.27 ± 0.04
GBM	0.27 ± 0.04	0.28 ± 0.04	0.27 ± 0.04
VOTE	0.25 ± 0.04	0.27 ± 0.04	0.25 ± 0.04

Table 9. Innolink Group

	Original	Features added	Added and selected
LDA	0.31 ± 0.01	0.24 ± 0.03	0.3 ± 0.03
KNN	0.34 ± 0.04	0.28 ± 0.04	0.28 ± 0.04
TREE	0.27 ± 0.03	0.27 ± 0.04	0.27 ± 0.03
SVM	0.29 ± 0.02	0.24 ± 0.03	0.26 ± 0.03
MNOM	0.34 ± 0.03	0.23 ± 0.03	0.3 ± 0.03
RF	0.19 ± 0.03	0.17 ± 0.03	0.18 ± 0.03
GBM	0.22 ± 0.03	0.19 ± 0.04	0.2 ± 0.04
VOTE	0.28 ± 0.02	0.23 ± 0.03	0.26 ± 0.03

Table 10. Papua merchandising

	Original	Features added	Added and selected
LDA	0.1 ± 0.03	0.11 ± 0.03	0.09 ± 0.03
KNN	0.12 ± 0.03	0.12 ± 0.03	0.09 ± 0.03
TREE	0.12 ± 0.06	0.12 ± 0.06	0.12 ± 0.06
SVM	0.12 ± 0.03	0.11 ± 0.03	0.1 ± 0.03
MNOM	0.09 ± 0.03	0.09 ± 0.03	0.08 ± 0.03
RF	0.1 ± 0.03	0.08 ± 0.03	0.07 ± 0.03
GBM	0.09 ± 0.03	0.08 ± 0.03	0.07 ± 0.03
VOTE	0.1 ± 0.03	0.08 ± 0.03	0.07 ± 0.03

Table 11. Lempesti

	Original	Features added	Added and selected
LDA	0.1 ± 0.03	0.12 ± 0.03	0.1 ± 0.03
KNN	0.12 ± 0.03	0.12 ± 0.03	0.07 ± 0.03
TREE	0.12 ± 0.06	0.12 ± 0.06	0.12 ± 0.06
SVM	0.12 ± 0.03	0.1 ± 0.03	0.08 ± 0.03
MNOM	0.09 ± 0.03	0.09 ± 0.03	0.08 ± 0.03
RF	0.1 ± 0.03	0.07 ± 0.03	0.07 ± 0.03
GBM	0.09 ± 0.03	0.08 ± 0.03	0.08 ± 0.03
VOTE	0.1 ± 0.03	0.07 ± 0.03	0.07 ± 0.03

Table 12. Toyota material handling Finland

	Original	Features added	Added and selected
LDA	0.36 ± 0.11	0.38 ± 0.12	0.32 ± 0.11
KNN	0.25 ± 0.08	0.2 ± 0.08	0.17 ± 0.08
TREE	0.39 ± 0.07	0.39 ± 0.07	0.39 ± 0.07
SVM	0.23 ± 0.08	0.2 ± 0.08	0.19 ± 0.07
RF	0.24 ± 0.08	0.24 ± 0.07	0.23 ± 0.07
VOTE	0.24 ± 0.08	0.21 ± 0.08	0.2 ± 0.08

In a previous preprocessing combinations study exactly the same Toyota and Innolink case datasets were used yielding VOTE misclassification error 0.16 for the former and 0.18 for the latter compared to 0.20 and 0.26 above. The execution time for preprocessing combinations was about 24 h with 16 vCPUs in the Amazon cloud contrast to approximately 10 min in the same setting for constructed data quality features. Programming the constructed features analysis with R statistical language took approximately 1/10 of time needed to program the preprocessing combinations approach. This was mostly because combinations of preprocessors require extensive exception and error handling capabilities.

6 Conclusions

Eight elementary forms of constructed featured were identified and operationalized.

From Tables 3, 4, 5, 6 and 7 can be observed that constructed data quality features can be among the most importance features in business performance measurement system prediction task.

From Tables 8, 9, 10, 11 and 12 comparing original (left columns) and mixed but equal dimensional sets (right columns) can be observed that adding constructed features performed better in 29/38 and equally in 9/39 tests of prediction performance as measured by classification accuracy. The constructed data quality features did not perform worse in the any of the tests.

Additionally, the hierarchical clustering of data quality features by using correlation as a measure of distance reveals that data quality features are dependent on data and vary among the example cases.

Contrasting to a competing method of preprocessing combinations with two datasets, constructed data quality feature performed slightly lower in accuracy, but clearly better in execution time and easiness of use measured as time needed to program the solution. Compared to preprocessing combinations constructed features are computationally more stable.

For research implications, the results suggest that constructed data quality features as input in predictive modelling of business performance measurement system data can improve predictive classification accuracy. This implies that there can be systematic patterns in the data production process that are associated with the target feature to be predicted.

For practice, the results suggest that incorporating data quality features into data can be a viable alternative to data preprocessing with additional benefit of visualizing non-random data quality patterns. Thus, constructed data quality features can be more effective as higher prediction performance, efficient as shorter execution time compared to data transformations and easy to use as lessened need to analyze, identify and transform data quality problems. Although constructed data quality features provide opportunities to visualize high dimensional data quality patterns, their prediction performance was slightly below that of preprocessing combinations.

There are limitations to the results. First, the mapping of constructed data quality features to data quality concepts is at best, tentative and LenghtOfIQR and Linear-Dependency contain untested assumptions. Secondly, due to the relatively small sample size, interesting classification approaches were tested but rejected. Wolpert's [42] stacking, which was incidentally the prime motivator to start the study in the first place, is a two-level classifier that aims to learn from the first level classifier results. Thus it was expected to find out that stacking would benefit most from constructed data quality features. However, with the given sample sizes especially in the Toyota case it was not possible to generate three uncontaminated samples with large enough sample sizes.

Further research is needed first, to expand the range of constructed data quality features and elaborate their operationalization from elementary to sophisticated ones. Secondly, to evaluate whether there are repetitive findings indicating that specific features perform continuously better than others. Thirdly, to create a conceptual framework for interpreting these kind of findings starting from the mapping of constructed data quality features to data quality concepts and ending in prescriptive rules concerning actions needed after identification of a data quality patterns.

Acknowledgements. Professor emeritus Pertti Järvinen, professor Martti Juhola and Dr. Kati Iltanen University of Tampere, Finland. After sales director Jarmo Laamanen Toyota Material Handling Finland, managing director Marko Kukkola Innolink Group, sales director Mika Karjalainen 3StepIt, managing director Olli Vaaranen Papua Merchandising and managing director Sirpa Kauppila Lempesti.

References

1. Abdul-Rahmana, S., Abu Bakara, A., Hussein, B., Zeti, A.: An intelligent data pre-processing of complex datasets. Intell. Data Anal. **16**, 305–325 (2012)
2. Bellman, R.E.: Dynamic Programming. Rand Corporation, Princeton University Press, New Jersey (1957)
3. Berthold, M.R., Borgelt, C., Höppner, F., Klawonn, F.: Guide to Intelligent Data Analysis – How to Intelligently Make Sense of Real Data. Springer, London (2010)
4. Breiman, L.: Random forests. Mach. Learn. **45**, 5–32 (2001)
5. Breunig, M.M., Kriegel, H.-P., Ng, R.T., Sander, J.: LOF: identifying density-based local outliers. In: Proceedings of ACM SIGMOD 2000 International Conference on Management of Data, pp. 93–104 (2000)
6. Caruana, R., Niculescu-Mizil, A., Crew, G., Ksikes, A.: Ensemble selection for libraries of models. In: Proceedings of ICML, p. 18 (2004)
7. Chandola, V., Banerjee, A., Kumar, V.: Anomaly detection: a survey. ACM Comput. Surv. **41**(3), 15 (2009)
8. Chapman, P., Clinton, J., Kerber, R., Khabaza, T., Reinartz, T., Shearer, C., Wirth, R.: Crisp-Dm 1.0 Step by Step Data Mining Guide. Crisp-DM Consortium (2000)
9. Crone, S.F., Lessmann, S., Stahlbock, R.: The impact of preprocessing on data mining: an evaluation of classifier sensitivity in direct marketing. Eur. J. Oper. Res. **173**(3), 781–800 (2005)
10. Engel, J., Gerretzen, J., Szymanka, E., Jeroen, J.J., Downey, G., Blanchet, L., Buydens, L.: Breaking with trends in preprocessing. TrAC Trends in Analytical Chemistry **50**, 96–106 (2013)

11. Fayyad, U., Piatetsky-Shapiro, G., Smyth, P.: The KDD process for extracting useful knowledge from volumes of data. Commun. ACM **39**(11), 27–34 (1996)
12. Filzmoser, P., Maronna, R., Werner, M.: Outlier identification in high dimensions. Comput. Stat. Data Anal. **52**(3), 1694–1711 (2008)
13. Franco-Santos, M., Kennerley, M., Micheli, P., Martinez, V., Mason, S., Marr, B., Gray, D., Neely, A.: Towards a definition of a business performance measurement system. Int. J. Oper. Prod. Manag. **27**(8), 784–801 (2007)
14. Freund, Y., Schapire, R.E.: A decision-theoretic generalization of on-line learning and an application to boosting. J. Comput. Syst. Sci. **55**(1), 119–139 (1995)
15. Guyon, I., Elisseeff, A.: An introduction to variable and feature selection. J. Mach. Learn. Res. **3**, 1157–1182 (2003)
16. Han, J., Kamber, M., Pei, J.: Data mining: Concepts and Techniques. Morgan Kaufmann, San Francisco (2012)
17. Hsu, C.-W., Chang, C.-C., Lin, C.-J.: A Practical Guide to Support Vector Classification. Taiwan National University, Taipei (2010)
18. Hodge, V.J., Austin, J.: A survey of outlier detection methodologies. Artif. Intell. Rev. **22** (2), 85–126 (2004)
19. Hu, M.-X., Salvucci, S.: A Study of Imputation Algorithms, Institure of Education Science, NCES, New York (1991)
20. Järvinen, P.: On Research Methods. Opinpajan kirja, Tampere (2012)
21. Kaplan, R.S., Norton, D.P.: the balanced scorecard – measures that drive performance. Harvard Bus. Rev. **71**(1), 71–79 (1992)
22. Kira, K., Rendell, L.A.: A practical approach to feature selection. In: Proceedings of the Ninth International Workshop on Machine Learning, pp. 249–256 (1992)
23. Kitchenham, B., Brereton, O.P., Budgen, D., Turner, M., Bailey, J., Linkman, S.: Systematic literature reviews in software engineering - a systematic literature review. J. Inf. Softw. Technol. **51**(1), 7–15 (2009)
24. Kriegel, H.-P., Borgwardt, K.M., Kröger, P., Pryakhin, A., Schubert, M., Zimek, A.: Future trends in data mining. Data Min. Knowl. Disc. **15**(1), 87–97 (2007)
25. Kriegel, H.-P., Kröger, P., Zimek, A.: Outlier detection techniqes. In: 16th ACM SIGKDD Conference on Knowledge Discovery and Data Mining, Washington, DC (2010)
26. Kuhn, M., Johnson, K.: Applied Predictive Modeling. Springer, New York (2013)
27. Kochanski, A., Perzyk, M., Klebczyk, M.: Knowledge in imperfect data in advances in knowledge representation. In: Ramirez, C. (ed), DOI: 10.5772/37714. http://www.intechopen.com/books/advances-inknowledge-representation/knowledge-in-imperfect-data (2012)
28. Kohavi, R., John, G.H.: Wrappers for feature subset selection. Artif. Intell. **97**, 273–324 (1997)
29. Kotsiantis, S.B., Kanellopoulos, D., Pintelas, P.E.: Data preprocessing for supervised learning. Int. J. Comput. Sci. **2**, 111–117 (2006)
30. Ludmila, K.: Combining Pattern Classifiers: Methods and Algorithms. Wiley-Interscience, New Jersey (2004)
31. Longadge, R., Dongre, S.S., Malik, L.: Class imbalance problem in data mining: review. Int. J. Comput. Sci. Netw. **2**(1) (2013)
32. March, S., Smith, G.: Design and natural science research on information technology. J. Decis. Support Syst. **15**(4), 251–266 (1995)
33. Nørreklit, H.: The balance on the balanced scorecard—a critical analysis of some of its assumptions. Manag. Acc. Res. **11**(1), 65–88 (2000)
34. Peltonen, J.: Dimensionality Reduction. Lecture Series, University of Tampere (2014)
35. Pyle, D.: Data Preparation for Data Mining. Morgan Kauffman, San Francisco (2003)

36. Quinlan, J.R.: C4.5: Programs for Machine Learning. Morgan Kauffman, San Francisco (1993)
37. Sadiq, S., Khodabandehloo, Y.N., Induska, M.: 20 Years of data quality research: themes, trends and synergies. In: ADC 2011 Proceedings of the Twenty-Second Australasian Database Conference, vol. 115, pp. 153–162 (2011)
38. Torgo, L.: Data Mining with R: Learning with Case Studies. CRC Press, Boca Raton (2010)
39. Vattulainen, M.: A method to improve the predictive power of a business performance measurement system by data preprocessing combinations: two cases in predictive classification of service sales volume from balanced data. In: Ghazawneh, A., Nørbjerg, J., Pries-Heje, J. (eds.) Proceedings of the 37th Information Systems Research Seminar in Scandinavia (IRIS 37), Ringsted, Denmark, pp.10–13 (2014)
40. Wu, X., Kumar, V., Quinlan, J.R., Ghosh, J., Yang, Q., Motoda, H., McLachlan, G., Ng, A., Liu, B., Yu, P.S., Zhou, Z.H., Steinbach, M., Hand, D.J., Steinberg, D.: Top 10 algorithms in data mining. Knowl. Inf. Syst. **14**(1), 1–37 (2008)
41. Wand, Y., Wang, R.: Anchoring data quality dimensions in ontological foundations. Commun. ACM **39**(11), 86–95 (1996)
42. Wu, X., Zhu, X., Wu, G.-Q., Ding, W.: Data mining with big data. IEEE Trans. Kowl. Disc. Data Eng. **26**(1), 97–107 (2013)
43. Wolpert, D.: Stacked generalization. Neural Netw. **5**, 241–259 (1992)
44. Yang, Q., Wu, X.: 10 Challenging problems in data mining research. Int. J. Inf. Technol. Decis. Mak. **5**(4), 597–604 (2006)
45. Zhao, H., Sudra, R.: Entity identification for heterogenous database integration —a multiple classifier system approach and empirical evaluation. Inf. Syst. **30**(2), 119–132 (2005)

How to Support Customer Segmentation with Useful Cluster Descriptions

Hans Friedrich Witschel$^{(\boxtimes)}$, Simon Loo, and Kaspar Riesen

University of Applied Sciences Northwestern Switzerland (FHNW),
Riggenbachstrasse 16, 4600 Olten, Switzerland
{HansFriedrich.Witschel,Kaspar.Riesen}@fhnw.ch,
Simon.Loo@students.fhnw.ch

Abstract. Customer or market segmentation is an important instrument for the optimisation of marketing strategies and product portfolios. Clustering is a popular data mining technique used to support such segmentation – it groups customers into segments that share certain demographic or behavioural characteristics. In this research, we explore several automatic approaches which support an important task that starts *after* the actual clustering, namely capturing and labeling the "essence" of segments. We conducted an empirical study by implementing several of these approaches, applying them to a data set of customer representations and studying the way our study participants interacted with the resulting cluster representations. Major goal of the present paper is to find out which approaches exhibit the greatest ease of understanding on the one hand and which of them lead to the most correct interpretation of cluster essence on the other hand. Our results indicate that using a learned decision tree model as a cluster representation provides both good ease of understanding and correctness of drawn conclusions.

1 Introduction

In order to optimise their marketing strategies, companies need to understand the needs and preferences of their customers closely. Customer segmentation is a technique that allows companies to group customers into segments that share certain characteristics such as preferences or demand [16]. Based on customer segments and an understanding of their meaning, product offerings and marketing strategies can be better targeted by distinguishing certain categories of needs.

Many companies have begun to see the potential in gaining competitive advantage by extracting knowledge out of the abundant data that they can collect about their customers' background, interests and behaviour. For instance, using data mining methods, they have been able to understand their customers' purchasing habits better. The prevalent data mining method for customer segmentation is clustering (see e.g. [12]). Clustering is used to divide objects – e.g. customers – in the data set into clusters (group of related data points) such that objects within a cluster all share certain similarities. As a basis for clustering, each data object needs to be described by certain *attributes* or *features*.

© Springer International Publishing Switzerland 2015
P. Perner (Ed.): ICDM 2015, LNAI 9165, pp. 17–31, 2015.
DOI: 10.1007/978-3-319-20910-4_2

The values of these attributes are compared between data objects to assess their similarity.

When segmenting customers of a company, an obvious and traditionally used family of attributes are demographic features, such as gender, nationality, family and socio-economic status. However, it has been recognised that attributes related to the interests and/or behaviour of customers can be more meaningful [13]. The resulting set of attributes usually has a mixture of types, including binary (e.g. interest in something, yes or no), categorical (e.g. nationality) and numeric attributes (e.g. number of times an event has occurred).

When clustering has been applied to a set of customers described by such attributes, the result is a set of clusters (or segments). For reasonably large customer bases, the size of such segments will be in the hundreds or thousands. Before a marketer can actually benefit from the result, (s)he needs to understand the "essence" of each segment, i.e. the characteristics that are shared by all customers within the segment and that make it different from the other segments. Usually, the marketer captures the essence of a segment by assigning a label, e.g. "rich singles who are active shoppers interested in accessories".

Given the aforementioned size of segments, capturing and labeling the essence of clusters is not an easy task and needs to be supported by providing some sort of automatically generated cluster descriptions. Current data mining tools offer only limited functionality: usually, they allow to visualise the univariate distribution of attribute values within clusters. The drawback of such functionalities is threefold. First, they fail to highlight which attributes are important for a cluster, second therefore tend to be quite tedious to analyse if there are many attributes and third, they do not capture multivariate effects, i.e. meaningful combinations of attributes.

Some research proposed alternative ways of describing the essence of clusters, but there has been – to the best of our knowledge – no systematic evaluation of the quality of such alternatives. Therefore, the goal of our research is to assess the quality of a selection of cluster description techniques – specifically for the application to customer segmentation – in terms of ease of understanding and correctness of drawn conclusions. That is, our primary question is "which description format will an end user find most efficient and effective in interpreting clustered data?" Here, we mean by "efficient" the time required by the end user to understand the description. By "effective" we mean how correctly the end user will understand the description. If a description leads to accurate interpretation of the clustered objects, then we can say that the description is "effective", if it leads to incorrect conclusions, it is ineffective. That is, we assume that on the one hand, marketers will prefer certain description methods because they are easy to grasp. On the other hand, a preferred method may be easy to grasp, but lead to wrong conclusions about the essence of a cluster, e.g. because of an oversimplification.

In order to carry out our evaluation, we proceeded as follows: we analysed the most promising proposed alternatives for segment descriptions from literature, as described in Sect. 2. We then derived a set of hypotheses regarding how accurately they describe cluster essence (or may lead to wrong conclusions, respectively), see Sect. 3. Subsequently, we designed an experiment to verify these hypotheses by selecting a data set of customers, clustering it with a standard clustering algorithm and representing it with the chosen alternative description methods. The hypotheses were translated into a questionnaire that was, together with the cluster descriptions, given to a number of test persons. The precise setup of the experiment is described in Sect. 4. We then analysed and coded the responses of participants and mapped out the results as described in Sect. 5. Finally, we were able to draw conclusions, see Sect. 6.

2   Related Work

Clustering is a very active area of research and a great variety of clustering algorithms exists – see [7] for an exhaustive overview. The general approach of most clustering algorithms is to first establish a measure of similarity between data objects and then to try to group them such that objects in the same cluster are maximally similar and objects belonging to different clusters are maximally dissimilar. As mentioned above, researchers have also applied clustering to the problem of customer segmentation in various ways and settings, e.g. [12,15].

The topic of how to describe and summarise the "essence" of clusters has received far less attention than the clustering of data itself. A notable exception is the area of text mining, where various methods for describing document clusters have been proposed (e.g. [3,10,14]). In document clustering, documents are usually represented by high-dimensional vectors where each term/word occurring in the whole document collection forms a dimension. For each dimension, a numerical attribute is created which represents the degree to which the corresponding term describes the content of the document. Hence, popular cluster description methods in text mining rely on the intuition that clusters should be represented by those terms that occur frequently within the cluster's documents, but rarely otherwise. A cluster description is then a set of terms.

When clustering customers,however, the situation is usually different: as explained in Sect. 1 above, customers are usually represented by a mixture of – comparatively few – binary, categorical and numerical attributes. For nominal attributes the intuition from the text mining area does not work.

The approaches to summarising clusters of structured data objects with mixed attribute types can roughly be distinguished into two directions:

- Approaches that **summarise the distribution of attribute values** within the cluster. Many data mining tools offer visualisation of such distributions. Often, however, this can also happen by exploiting summaries that are built into existing clustering algorithms and that are simpler and faster to inspect by a human. For instance, the popular k-means algorithm [9] uses so-called *centroids* of clusters. A centroid is a vector of attribute values where each value summarises the distribution of values of a given attribute for all cluster members. For numerical attributes, the centroid contains the arithmetic mean of all values, for categorical attributes, it contains the mode.

Other clustering approaches use more verbose summaries of distributions, especially for categorical attributes. For instance, the COBWEB algorithm [4] – an instance of so-called *conceptual clusterers* – represents clusters by a set of conditional probabilities, namely $P(A_i = V_{ij}|C_k)$ where A_i is a categorical attribute, V_{ij} represents one of the values of attribute A_i and C_k is a cluster. This essentially maps out all frequencies of the values of a categorical attribute within a cluster. A similar representation – using plain frequencies instead of probabilities – can be obtained for the expectation maximisation (EM) algorithm [2], a fuzzy clustering algorithm of which k-means is a special case.

– The other class of approaches relies on **learning a classifier with human-interpretable model** that is able to distinguish between the induced clusters. For instance, in [5,6], it is proposed to learn a decision tree from the clustered data. This means that one first clusters the data and then trains a decision tree classifier to predict the cluster for unknown data objects, i.e. using the cluster number of each data object as class attribute. The resulting decision tree can then be inspected by a human. An important characteristic of a decision tree is the way in which it arranges attributes: given a decision tree for cluster C_k classifying objects into either "C_k" or "*not* C_k", the top-most attribute of that tree is the one that contributes most to reducing the uncertainty about whether an object belongs to C_k or not. This means that usually the top-most attribute is the one that most captures the "essence" of the cluster. If a tree becomes too large to be inspected easily by a human, it can be *pruned* (see e.g. [11]) such that only the most important attributes are visible.

Although there is no explicit proposal in the literature, other classifiers with human-interpretable models could be used in the same way. For instance – as can be seen e.g. by the comparative evaluation of classifiers in [8] – rule learners also yield models that humans can easily understand, e.g. RIPPER [1]. In that case, the model consists of a set of interdependent rules of the form *if $A_i = V_{ij}$ and $A_l = V_{lm}$ and ... then C_k* (see Sect. 4 for an example).

Although these possibilities have been proposed in literature and some of them are surely used in practice today, there has been no systematic and empirical evaluation of the suitability of these approaches to the problem of capturing and labeling the essence of clusters. We have chosen to contrast k-means centroids – as a member of the first category of approaches – with decision tree and rule representation – as representatives of the second category. We are aware that more sophisticated approaches exist for the first category – but we found it important to evaluate centroids because of their popularity, and chose not to evaluate the other ones because of their greater complexity and the limited number of participants that we could recruit.

3   Hypotheses

Taking into account the different characteristics of the representation approaches described in the previous section, it is natural to assume that these characteristics will have an impact on the correctness of conclusions that a human draws when inspecting the representations. In the following, we will discuss our expectations regarding that impact for our chosen representations (centroid, decision tree and rules, see last section) and derive hypotheses, to be tested in the empirical part of our work.

3.1   Centroid Representation

Analysing the characteristics of the centroid representation leads to the following assumptions: First, we note that a centroid summarises *numerical attributes* by arithmetic mean of all values within the cluster. To a person with a background in statistics, it is clear that the values of the cluster elements are not necessarily all close to the mean. There can be various reasons for this – for instance, there could be an outlier that is pulling the mean or the variance of this particular attribute could be large. However, we assume that marketers may not be very educated in statistics and that even persons who do have a reasonable education in statistics, might be (mis-)led by the arithmetic mean to believe that the mean is representative of a majority of the attribute distribution's values. For example, when analysing an attribute such as the age of persons within a cluster, they will believe that most people in the cluster have an age close to the arithmetic mean. We phrase this as the following hypothesis:

H1: Given a cluster representation that summarises numerical attributes using the arithmetic mean of the attribute values in the cluster, a human analyst will be led to believe that most of the attribute values are close to the mean and hence believe that the cluster can be characterised by the mean value w.r.t. the given attribute.

Second, we note that a centroid summarises *categorical attributes* by the mode, i.e. the most frequent of all values of the attribute within the cluster. Now, if there are two or more values of the attribute's value distribution with almost equal frequency within the cluster, this will not be realised. As an example, consider analysing the gender of persons. If 51 % of persons in a cluster are female and 49 % are male, the mode of the gender attribute will be female. However, it is wrong to conclude that the cluster consists primarily of females (the number of females being not much higher than that of males and equal to our overall expectation). From this, we derive another hypothesis as follows:

H2: Given a cluster representation that summarises categorical attributes using the mode of the attribute values in the cluster, a human analyst will be led to believe that the vast majority of cluster members has that attribute value and hence believe that the cluster can be characterised by the mode w.r.t. the given attribute.

3.2   Rule Representation

When analysing rules, one needs to be aware of two important facts: First, each rule only captures the characteristics of *some* cluster members and one needs to unite all statements made by the rules in order to capture a comprehensive picture of the cluster. Second, all statements about attribute values made within a rule need to be taken together, i.e. interpreted in a conjunctive way. Since this is a rather complex matter, we may assume that even if we instruct human analysts about it, there is a risk that they take statements from a single rule in isolation and generalise them to the whole cluster.

> *H3: Given a rule cluster representation and an explicit question concerning an attribute-value combination that is used within only one or two rules and that is not necessarily predominant in the whole cluster, a human analyst will state that in fact the combination is representative for all cluster members.*

3.3   Decision Tree Representation

An important characteristic of a decision tree is that the attributes that are on top of the tree are the ones that contribute most to reducing the uncertainty about cluster membership, i.e. that usually the top attributes are the ones that most capture the "essence" of a cluster. We assume that this characteristic will naturally lead them to using attributes from the top of the tree when asked to provide a label for a cluster. At the same time, since a centroid representation does not provide any information about importance of attributes, we may assume that human analysts will choose attributes randomly or according to the order in which they are presented in the centroid. For rule representations, we may assume that an analyst will use those attributes that appear most frequently in the whole set of rules. These observations bring us to the following fourth hypothesis:

> *H4: When asked to provide a label to describe the essence of a cluster, a human analyst using*
>
> a. *a decision tree representation of the cluster will use attributes from the top of that tree in the label*
> b. *a centroid representation will choose attributes randomly to formulate the label*
> c. *a rule representation will use attributes in the label that appear most frequently within the rules*

Another assumption is that, when we explicitly ask a human analyst for the importance of an attribute that is at a low level of a decision tree representation, there is a risk that (s)he forgets about the instructions (see Sect. 4.2) and confirms that importance:

> *H5: Given a decision tree cluster representation and the question of whether an attribute at a low level of the tree is important for the essence of the cluster, a human analyst will be misled to say that the attribute is important.*

The above hypotheses all aim at assessing the "effectiveness" aspect of representation, i.e. whether or not they lead a marketer to the right conclusions regarding the essence of a cluster. Regarding the efficiency aspect, i.e. the question how easy and fast marketers can derive the cluster essence using a particular representation, we need to formulate an open question (since we do not have an a priori assumption):

> *Q1: Given a set of alternative representations for a cluster, which representation will human analysts prefer? That is, for which representation will they state that it is easiest to derive the essence of a cluster?*

And finally, we are interested to learn about the reasons for such preference:

> *Q2: Given the preference of a human analyst for a particular type of cluster representation, which reasons, i.e. which specific characteristics of the representation, lead to this preference?*

4   Experimental Setup

4.1   Data Selection and Preparation

For our empirical validation of the hypotheses, we first needed a data set comprising description of persons by demographic and behavioural attributes. We chose the data of the 1998 KDD Cup[1], a data set that profiles persons, using 479 different attributes, with the purpose of predicting their reaction to a donation-raising campaign.

With that data, we proceeded as follows:

- We drew a random sample of 8000 persons
- We made a selection of 35 attributes, comprising 6 demographic attributes (combined socio-economic status and urbanicity level, age, home ownership, number of children, income and gender), 14 behavioural, numerical attributes indicating the number of known times a person has responded to other types of mail order offers and 15 binary variables reflecting donor interests (with values "yes" or "no"), as collected from third-party data sources.
- We clustered the data using k-means, setting the number of clusters to 15, and recorded the resulting centroids.
- We selected two clusters (cluster 5 and cluster 13) from that result.
- We built a decision tree (C4.5) and RIPPER rule model to distinguish between members of the two selected clusters and the rest of the data set, respectively.
- We represented the two decision trees in a user-friendly graphical form.

Thus, we had, for each of the two selected clusters, a centroid representation in a tabular format, a graphically represented tree and a set of text-based rules. Figure 1 shows the centroid representation of the two clusters – along with the mean and mode values for the full data set (which can be used for comparison and to detect potentially meaningful deviations). Figure 2 shows part of the tree representation and the full rule representation of cluster 5.

---

[1] http://www.sigkdd.org/kdd-cup-1998-direct-marketing-profit-optimization.

Attribute	Full Data	Cluster 5	Cluster 13
# persons	8000	79	818
neighbourhood	rural middle-class	urban upper-class	town middle-class
age	62.6855	53.1454	70.848
home owner	Y	Y	N
# children	1.5157	1.5125	1.5052
household income	3915.9	4303.8	2152.8
gender	F	F	F
Buy Craft articles	0	0	0
buy gardening articles	0	0	0
buy books	1	0	0
buy collectables	0	0	0
react to health pubs	0	0	0
buy male magazines	0	1	0
...	...	...	...
interest in collectables			
interest in veteran topics	Y		
interest in bible reading			
interest in catalog shopping	Y		
...	...	...	...

Fig. 1. Centroid representation (extract) of the two clusters

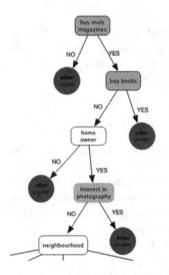

- **If** (BUY MALE MAGAZINE = 1) and (BUY BOOKS = 0) and (REACT TO HEALTH PUBS = 0) **then Cluster 5**
- **If** (BUY MALE MAGAZINE = 1) and (BUY BOOKS = 0) and (AGE <= 52) **then Cluster 5**
- **If** (BUY MALE MAGAZINE = 1) and (BUY BOOKS = 0) and (INCOME <= 4) and (BUY FAMILY MAGAZINES = 0) **then Cluster 5**
- **If** (BUY MALE MAGAZINE = 1) and (GENDER = F) and (AGE <= 48) and (BUY FAMILY MAGAZINES = 0) and (INCOME >= 4000) **then Cluster 5**
- **If** (BUY MALE MAGAZINE = 1) and (REACT TO HEALTH PUBS = 0) and (REACT TO CULINARY PUBS = 1) **then Cluster 5**
- **If** (URBAN UPPER-CLASS) and (AGE <= 45) and (BUY BOOKS = 0) **then Cluster 5**
- **If** (URBAN UPPER-CLASS) and (BUY BOOKS = 0) and (GENDER = F) and (INTEREST IN PETS = Y) **then Cluster 5**

Fig. 2. Tree (extract) and rule representation of cluster 5

4.2 Participants and Questionnaire

Next, we prepared a set of tasks and questions to test the hypotheses presented in Sect. 3. We recruited a total of 46 participants who answered the questionnaire. We divided the participants into three groups, one for each type of representation, resulting in groups that we will call "centroid group", "tree group" and "rules group". Both the centroid and tree group had 15 members, the rules group had 16. All participants were students at our school. Hence, all of them

have come across the topic of statistics within their studies, i.e. should be familiar
with statistical basics.

Each group received the same questionnaire, plus one sheet of paper containing the group-specific representation of both cluster 5 and 13 (e.g. a tree for
the tree group). In addition, each participant was given a folded sheet containing the other two representations (e.g. rules and centroid for the tree group) of
cluster 5. Participants were asked to leave that sheet folded, i.e. not to look at
it before they were asked to. Finally, participants were instructed, both orally
and in written form, about the meaning of attributes and the meaning of representations, e.g. it was mentioned that centroids contain modes and means, that
decision trees have important attributes at the top and that rules may describe
only a subset of a cluster.

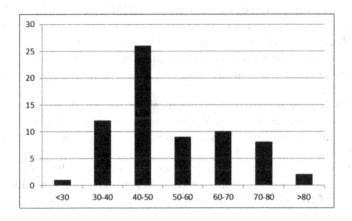

Fig. 3. Distribution of the age attribute in cluster 5

In the following, we report the questions of our questionnaire, along with our
expectations regarding how each group of participants will react:

Task 1. To test hypothesis $H1$, we asked participants to look at their representation of cluster 5 and answer the question "What age group do you think
most representative of this cluster?" Fig. 3 shows the age distribution within
cluster 5 as a histogram – clearly the age group between 50 and 60 is not predominant in this cluster, as one could believe by analysing the mean value,
53.15 in the centroid (see Fig. 1), but we expect the centroid group to fall
into that trap according to $H1$.

Task 2. For $H2$, we asked participants to look at their representation of cluster 13 and respond to the question "How would you make a generalisation
about the demographic of this cluster by analysing the variables *neighbourhood, age, household income and gender*?". Here, we are only interested in
the "neighbourhood" attribute. When closely analysing its distribution, one
finds that cluster members are not very rich (only 9 % have the value "upper-
class"). The other part of the attribute does not show a clear picture – the

mode of the attribute, which indicates that cluster members live in towns, only applies to 37 % of the cluster members. Hence, it is wrong to say that cluster members live in towns. Since the tree does not contain the neighbourhood attribute, we expect the tree group to indicate that no answer is possible and, according to $H2$, the centroid group to "fall into the trap" of saying "towns". For the rule group – where various values of the attribute with various urbanicity levels are mentioned – we expect that only the socioeconomic status ("middle or lower class") will be mentioned by participants, but not the urbanicity level.

Task 3. To test $H3$, we let participants look at their representation of cluster 5 and then asked "Is there any indication of the health consciousness of this cluster and how would you describe it?" In cluster 5, the reaction to health pubs is more frequent (47 %) than in the full data (20 %). However, since two rules for cluster 5 contain the statement "REACT TO HEALTH PUBS = 0", and according to $H3$, we expect that the rule group will say that cluster 5 members are predominantly not health conscious. For the other groups, we expect that they do not see an indication since the tree does not contain the attribute and the centroid does not report a value that deviates from the full data.

Task 4. For $H4$, we asked participants to write down a short label to characterise the essence of cluster 5. We then analysed which attributes they had used in their labels and compared them to our expectations according to hypotheses $H4$, a), b) and c).

Task 5. To test $H5$, we asked participants to infer, from their representation of cluster 13, an answer to the question "Is pet ownership a significant characteristic of this cluster?" According to $H5$, we expected all groups to deny this importance.

Task 6. Finally, we told participants to open the folded sheet containing the alternative representations of cluster 5 and then asked the question "Would you have found it easier to label the clusters with one of these alternative representations? If so, which of the two?" From the answers, we collected evidence to answer our research questions $Q1$ and $Q2$.

5 Results and Discussion

Since all questions that we gave to the participants were to be answered freely, i.e. not using multiple choice, their answers had to be coded before a quantitative analysis became possible. Below, we report such analysis and discuss the results, at the same time explaining the codes that we formed inductively while reading the answers. Sometimes, coding required interpretation, especially when analysing the cluster labels that participants were asked to write down in Task 4. For instance, the tree representation of cluster 5 does not contain the gender attribute. But it does contain the attribute "buy male magazines" at the top of the tree (see Fig. 2). Hence, when participants stated that cluster members were male, we assumed that they used the "buy male magazines" attribute to infer that.

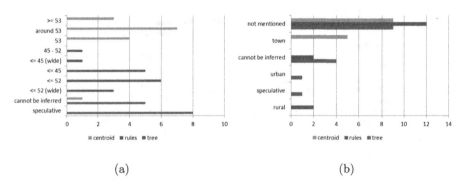

(a)   (b)

Fig. 4. Frequency of answers (coded) for Tasks 1 and 2 (colour in on-line proceedings)

Figure 4(a) shows the coded answers and their frequency of occurrence for Task 1. Here and henceforth, the code *cannot be inferred* means that a participant made an explicit statement that there is no or not enough evidence to infer an answer to the task. We generally use the code *speculative* to summarise all answers where the participant gave an answer that did not refer to values of the attribute in question, but where it was obvious that this value was inferred from other attributes. For instance, a member of the tree group gave the answer "15 to 25 (income less than 4000) or 65 to 80 (income lower, home owners, collectors, gardening)" for Task 1, which shows that age was inferred from income and some interest variables. The codes ≤ 45 *(wide)* or ≤ 52 *(wide)* summarise all answers where the participants indicated that *all* ages below 45 and 52 respectively are included. The frequencies of answers show very clearly that all members of the centroid group fell into our trap and derived an answer from the (misleading) mean. Only one centroid group member indicated that "cluster 5 only has the average age", which we coded as *cannot be inferred*. Members of the tree group either speculated or indicated that age cannot be inferred since the age attribute is not present in the tree. Members of the rule group made statements that better reflect the actual age distribution as shown in Fig. 3. In summary, we conclude that we can clearly accept hypothesis $H1$.

Figure 4(b) shows coded answers and their frequency for Task 2. We only coded the urbanicity level part of the answer, which was sometimes not mentioned in the answer (resulting in the code *not mentioned*). The codes *cannot be inferred* and *speculative* are defined as above. We can see that, if members of the centroid group mention the urbanicity level, they say "town". Only three members of both other groups mention the urbanicity level at all, consistent with our expectations. Thus, although the support for this is smaller than expected (with only 5 explicit answers from the centroid group), we can carefully confirm $H2$.

Next we look at the result of Task 3, displayed in Fig. 5(a). Here, the codes *yes, positive* and *yes, negative* denote that participants said that there is an indication of health consciousness and that it is positive or negative, respectively. The code *yes, undirected* means that the participants only said there is indication of health consciousness but did not indicate anything from which to conclude the

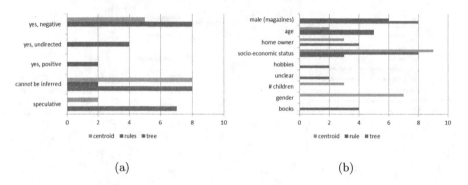

(a)   (b)

Fig. 5. Frequency of answers (coded) for Tasks 3 and 4 (colour in on-line proceedings)

direction. The results show that, as expected, 8 members of the rule group fall into the trap of concluding, from statements embedded in only two rules, that cluster 5 members are not health conscious. We can assume that if we insisted on an answer to the direction question, some of those 4 rule group members who indicated no direction would also choose a negative health consciousness. Hence, we can conclude that there is at least a rather large danger of such conclusion for human analysts with a rule representation and thus carefully confirm $H3$. The danger also seems to exist – to a smaller extent – for the centroid group: although the mode here does not deviate from the full data, some people are led to believe that health consciousness is negative, just because the value is 0 for this attribute.

The results of Task 4 are displayed in Fig. 5(b). Besides the names of attributes, we have used the codes *unclear*, which means that no direct reference to an attribute could be detected in the answer and *hobbies* which describes answers that refer to all interest attributes as a whole. Here, the sum of frequencies of codes across each group are larger than the group size since many participants used more than one attribute in their answer. The results show that the different groups use rather different attributes for labeling. We see that there is a strong tendency for members of the tree group to use attributes from the top of the tree, i.e. mainly "buy male magazines" and "home owner". It is a bit surprising that "buy books" is not used although at second-highest position in the tree and that many participants in this group use age which can only be speculative since the tree does not contain the age attribute. We may assume that having carried out Task 1 before had an influence here – especially since the same people who speculated in Task 1 did the same here and their answers were consistent between Tasks 1 and 4. When inspecting the answers from the centroid group, we see that participants used exclusively demographic attributes to describe the cluster – which may be explained by the fact that they are at the top of the centroid representation (see Fig. 1). Their choice is hence, although not completely random, not motivated by any meaningful criteria. Members of the rule group use primarily the attributes *socio-economic status, buy male magazines*, *age* and *buy books*. Except age, all of these are frequent or mid-frequent in the rules for cluster 5, see Fig. 2. All of these observations, taken together, let us rather clearly accept *H*4.

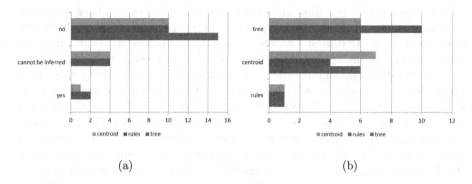

(a)   (b)

Fig. 6. Frequency of answers (coded) for Tasks 5 and 6 (colour in on-line proceedings)

Figure 6(a) shows answers for Task 5. We can see that *H*5 can be rejected very quickly since no member of the tree group found pet ownership to be a significant characteristic of cluster 13. Unexpectedly, a total of three participants from both the centroid and rules group think that this is the case – for unknown reasons.

Finally, we see the preferences of participants regarding the representation that allows labeling with greatest ease in Fig. 6(b). We see that rules are unpopular in all groups. Trees and centroids are equally popular in the tree and centroid groups. Participants from the rules group show a stronger preference for the tree such that this is also the overall most popular choice. Finally, we have coded the reasons that participants gave to explain their preference (if any). The code *simplicity* refers to answers that said that a representation is easy to understand/grasp. *Completeness* means that participants liked the fact that a representation gives overview over all (or enough) attributes. The attribute *comparison* refers to the ease of comparing several cluster, *attribute order* to the ease of identifying important attributes easily and *precision* to the space that a representation leaves for false interpretations.

Arguments in favour of the tree representation were mainly simplicity (7 mentions), as well as attribute order and precision (1 mention each). Rules – if preferred – were liked for their *completeness* (2 mentions). The centroids are also preferred because of *completeness* (3 mentions), and additionally for their *simplicity* and possibility of *comparison* (1 mention each). To sum up, the main reason for choosing a tree – according to our participants – is *simplicity*, whereas a possible reason for choosing centroids could be their *completeness*.

6   Conclusions and Future Work

In our experiments, we were able to confirm all of our intuitions about the process of capturing cluster essence and possible false conclusions that may result. In summary, we saw that centroids have several severe weaknesses, both in the sense that it is hard to identify the attributes that most contribute to cluster essence and that false conclusions may result from looking only at mean or mode values – even for humans who have a moderate background in statistics.

Decision trees do not show such undesired characteristics and are also most popular with our participants when it comes to the question of how easily cluster essence can be inferred from a representation. Our experiment has also revealed additional interesting arguments in favour of certain representations, which need to be assessed in light of our other results as follows: in light of the traps that people fell into, the only really valid argument in favour of centroids is their strength in allowing to easily compare several clusters. Trees score in simplicity. Attribute order – again considering the confirmation of our hypotheses – should also be seen as an important argument in favour of decision trees since it results in labels that better reflect truly important attributes. In this context, completeness should not be counted as a valid argument (used in favour of centroids or rules) since it does not serve a meaningful purpose to show summaries of attributes that do not contribute to describing the essence of a cluster – and may only lead to false conclusions about that essence.

For future research, it will be interesting to investigate more possible cluster representations, such as ones resulting from the application of expectation maximisation or COBWEB clustering – and to develop and test new hypotheses that go along with those representations. This includes also advanced centroid representations that use e.g. confidence intervals for means, try to order attributes by importance or show full distributions of categorical attributes. Similarly, future work might want to study improved rule representations, e.g. by showing the coverage of rules or play with different levels of pruning the decision trees. In addition, to get a deeper understanding, one might perform separate in-depth analyses for each kind of attribute type, study different kinds of distributions of numerical attributes and analyse the effects of visualisation techniques.

References

1. Cohen, W.: Fast effective rule induction. In: Proceedings of the Twelfth International Conference on Machine Learning, pp. 115–123 (1995)
2. Dempster, A., Laird, N., Rubin, D.: Maximum likelihood from incomplete data via the EM algorithm. J. R. Stat. Soc. Ser. B (Methodol.) **39**, 1–38 (1977)
3. Radev, D.R., Jing, H., Stys, M., Tam, D.: Centroid-based summarization of multiple documents. Inf. Process. Manage. **40**(6), 919–938 (2004)
4. Fisher, D.: Knowledge acquisition via incremental conceptual clustering. Mach. Learn. **2**(2), 139–172 (1987)
5. Gordon, A.: Classification, 2nd edn. Taylor and Francis, London (1999)
6. Huang, Z.: Clustering large data sets with mixed numeric and categorical values. In: Proceedings of the First Pacific-Asia Conference on Knowledge Discovery and Data Mining, pp. 21–34 (1997)
7. Jain, A.K.: Data clustering: 50 years beyond K-means. Pattern Recogn. Lett. **31**(8), 651–666 (2010)

8. Kotsiantis, S.: Supervised machine learning: a review of classification techniques. Informatica **31**(3), 249–268 (2007)

9. MacQueen, J.: Some methods for classification and analysis of multivariate observations. In: Proceedings of the Fifth Berkeley Symposium on Mathematical Statistics and Probability, vol. 1, pp. 281–297 (1967)

10. Popescul, A., Ungar, L.: Automatic labeling of document clusters (2000), http://citeseer.nj.nec.com/popescul00automatic.html

11. Quinlan, R.: C4.5: Programs for Machine Learning. Morgan Kaufmann Publishers, San Mateo (1993)

12. Saglam, B., Salman, F., Sayin, S., Trkay, M.: A mixed-integer programming approach to the clustering problem with an application in customer segmentation. Eur. J. Oper. Res. **173**(3), 866–879 (2006)

13. Teichert, T., Shehu, E., von Wartburg, I.: Customer segmentation revisited: The case of the airline industry. Transp. Res. Part A: Policy Pract. **42**(1), 227–242 (2008)

14. Treeratpituk, P., Callan, J.: Automatically labeling hierarchical clusters. In: Proceedings of the 2006 International Conference on Digital Government Research, pp. 167–176 (2006)

15. Weber, R.: Customer segmentation for banks and insurance groups with fuzzy clustering techniques. In: Fuzzy Logic, pp. 187–196 (1996)

16. Wedel, M., Kamakura, W.: Market Segmentation: Conceptual and Methodological Foundations, 2nd edn. Kluwer Academic Publishers, Boston (2000)

Retail Store Segmentation
for Target Marketing

Emrah Bilgic[1]([✉]), Mehmed Kantardzic[2], and Ozgur Cakir[1]

[1] Faculty of Business Administration, Marmara University, Istanbul, Turkey
{emrah.bilgic,o.cakir}@marmara.edu.tr
[2] Speed School of Engineering, University of Louisville, Louisville, USA
mmkant01@louisville.edu

Abstract. In this paper, we use Data Mining techniques such as clustering and association rules, for the purpose of target marketing strategy. Our goal is to develop a methodology for retailers on how to segment their stores based on multiple data sources and how to create marketing strategies for each segment rather than mass marketing. We have analyzed a supermarket chain company, which has 73 stores located in the Istanbul area in Turkey. First, stores are segmented in 5 clusters using a hierarchical clustering method and then association rules are applied for each cluster.

Keywords: Clustering · Association rules · Market basket analysis · Segmentation

1 Introduction

The issue of customer segmentation for retail companies has been studied in many papers due to the fact that target marketing rather than mass marketing has become an obligatory strategy after the era of intense competition. That is why retailers have started to focus on customer needs, past shopping behavior and as a result localized some of the assortments. One of the challenges for supermarket chain companies operating different stores in different regions is how to create marketing strategies for each store, especially if the company does not have enough information about the demographics of its customers.

A business practice, Customer Relationship Management (CRM), is a combination of people, processes and technologies [1]. It seeks to understand and build long term profitable relationships with customers and markets [2]. In most of the researches that purposed to segment the customers for CRM, some demographics of the customers were already known to researchers through the company's loyalty cards. Demographics of the customers are always crucial for CRM tasks since the desired information cannot be reached by transaction records only. However, what if a supermarket chain company does not have enough information about the demographics of its customers? What other variables than customer demographics and transaction data can lead retailers to differentiate their stores for the sake of successful target marketing?

Store segmentation can be helpful for this issue. It aims to divide a network of stores into meaningful groups. For example, a retailer with a network of 350 stores may

© Springer International Publishing Switzerland 2015
P. Perner (Ed.): ICDM 2015, LNAI 9165, pp. 32–44, 2015.
DOI: 10.1007/978-3-319-20910-4_3

generate 6 different store types, with each comprised of a 'similar' set of stores based on user-defined variables [3]. "In the language of the economist, segmentation is disaggregative in its effects and tends to bring about recognition of several demand schedules where only one was recognized before" [4], furthermore market segmentation, one of the most fundamental strategic marketing concepts, is the process of dividing the potential customers into homogeneous subgroups in order to create target marketing. This segmentation process necessitates a complete analysis of the entire market, not only customer's needs and shopping habits but also knowledge of changing market conditions and competition [5].

Appropriate data mining tools, which are suitable for identifying useful information from the company's databases, are one of the best supporting tools for CRM decisions [6]. Cluster analysis, an appropriate data mining tool for segmentation, identifies homogenous groups of objects that, objects in a specific cluster have similar characteristics but have dissimilar characteristic than the objects not belonging to that cluster. The problem, facing segmentation studies is to decide what characteristics to use for segmentation purpose, in other words which clustering variables will be included in the analysis. Past researches proved that some characteristics such as socioeconomic characteristics of the trade area, competition, sales area and special services that stores offer affect the performance of stores, which may lead us to find similar and dissimilar stores in terms of those characteristics [5, 7–9].

After segmenting stores using clustering, frequent itemsets and association rules analysis will be applied for each cluster for the purpose of target marketing. This way, we will be able to find the differences between the clusters in terms of the items sold at the stores.

2   Retail Store Segmentation Based on Multiple Data Sources

In this paper, the variables for clustering task were collected from different sources as shown in Table 1 below. Furthermore, Figs. 1, 2 and 3 summarize these variables for the data at hand. Variables related to each store are: store size, whether the store is closer to a university, a factory, a trade center and a touristic place or not, whether the store has a special parking lot or not, whether the store has a bus service for customers or not, and finally the number of competitor stores nearby.

Because the retailer company does not have information about its customers for each store, data about the trade area demographics were provided by the Turkish Statistical Institute. The people are considered as potential customers of each store since the data is exactly for the districts where 73 stores are located. Data includes the distribution of the population; by age groups, by marital status and by completed education level for the people residing in districts where the stores located. Average apartment rentals nearby those 73 stores are also found from several real estate web sites, in order to estimate the wealth of the people residing around the stores.

We removed highly correlated variables from our data set to prevent giving more weight to correlated data while calculating the dissimilarity matrix. There are high correlations found between Age 0–19 and Age 40–59, Age 0–19 and Age 60+, High Education and Low Education variables. Percentage of Single and Married variables

Table 1. Type, scale and source of variables used for the segmentation task

Variable	Type	Scale	Source
Store size	Numeric	Square meter	Retailer's web site
Average rental	Numeric	Turkish Currency	Real estate companies
Competitor	Numeric	Number	Google Maps
Single	Numeric	Percentage	Turkish Statistical Institute
Married	Numeric	Percentage	Turkish Statistical Institute
Age 0–19	Numeric	Percentage	Turkish Statistical Institute
Age 20–39	Numeric	Percentage	Turkish Statistical Institute
Age 40–59	Numeric	Percentage	Turkish Statistical Institute
Age 60+	Numeric	Percentage	Turkish Statistical Institute
Low educated	Numeric	Percentage	Turkish Statistical Institute
Middle educated	Numeric	Percentage	Turkish Statistical Institute
High educated	Numeric	Percentage	Turkish Statistical Institute
Factory area	Binary	0–1	Google Maps
University area	Binary	0–1	Google Maps
Trade area	Binary	0–1	Google Maps
Touristic area	Binary	0–1	Google Maps
Car park	Binary	0–1	Retailer's web site
Bus service	Binary	0–1	Retailer's web site

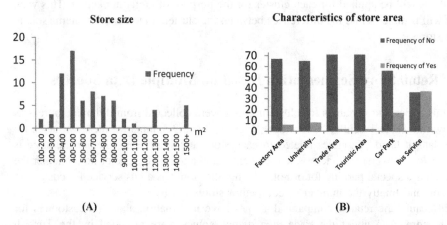

(A)   (B)

Fig. 1. Data representation; **(A)** Distribution of store sizes **(B)** Indicates number of stores which are located in factory area, university area…, have car parking, bus service or not

have correlation of −1. So we removed Age 0–19, High Education and Single variables before clustering analysis.

2.1   Methodology for Segmentation

Cluster analysis, the art of finding groups in data, is an unsupervised classification of objects into groups. It identifies homogenous groups of objects that, objects in a

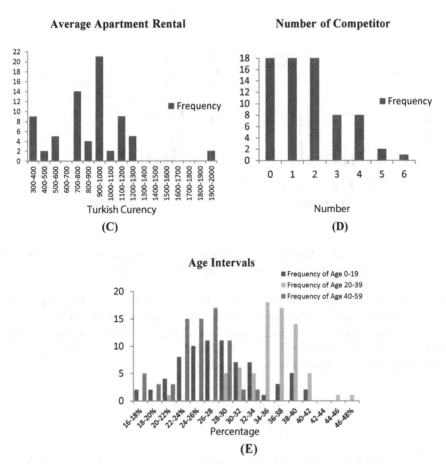

Fig. 2. Data Representation; (**C**) Distribution of average apartment rentals nearby 73 stores (**D**) Distribution of competitor stores nearby (**E**) Distribution of age intervals for the people, considered as potential customers, residing nearby 73 stores

specific cluster have similar characteristics but have dissimilar characteristic from the objects not belonging to that cluster [10]. Recent trends in retailing have shown that, retailers should specialize on target marketing by market segmentation rather than product specialization and mass marketing [5]. Below we provide the details of the clustering method we used for segmentation of the stores.

2.1.1 Dissimilarity Measure

Our data set derived from 73 stores, comprises mixed type of variables; 6 binary and 12 numeric variables. To be able to implement cluster analysis, first we need to measure the proximity between objects; in our case the stores, across each variable. Proximity is usually measured by a distance function.

Calculating the distances between objects of non-continuous variables is problematic. Gower has developed a proximity measure for mixed type variables [11]. It combines the different variables into a single proximity matrix. The definition of Gower takes

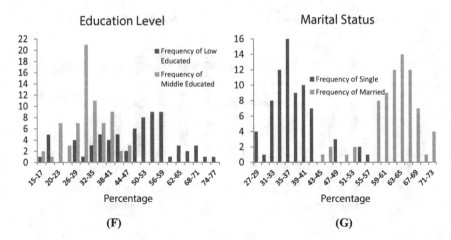

Fig. 3. Data representation; distribution of **(F)** Completed education level **(G)** Marital status, for the people, considered as potential customers residing nearby 73 stores

care of interval, nominal and binary data. Kaufman and Rousseeuw, generalized Gower's distance method which also covers ordinal and ratio variables [12].

They defined how to handle mixed types of variables for the dissimilarity matrix as follows; Let the data set contain p variables which are in different types. Then, the dissimilarity between objects i and j is defined as;

$$d(i,j) = \frac{\sum_{f=1}^{p} \delta_{ij}^{(f)} d_{ij}^{(f)}}{\sum_{f=1}^{p} \delta_{ij}^{(f)}} \tag{1}$$

$\delta_{ij}^{(f)}$ is equal to 1 when both measurements x_{if} and x_{if} for the fth variable are nonmissing and 0 otherwise. $d_{ij}^{(f)}$ is the contribution of the fth variable to the dissimilarity between i and j.

If variable f is interval-scaled, then $d_{ij}^{(f)}$ is given by

$$d_{ij}^{(f)} = \frac{|x_{if} - x_{jf}|}{R_f} \tag{2}$$

where R_f is the range of variable f.

If variable f is asymmetric binary scaled, then Jaccard Coefficient (1908) is appropriate for calculating the dissimilarity matrix;

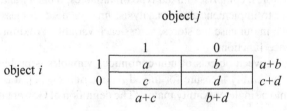

$$d_{ij}^{(f)} = \frac{b+c}{a+b+c} \tag{3}$$

After computing these formulas according to the variable types in the final, we have a 73 by 73 dissimilarity matrix. The elements of this matrix obtained by Formula 2 will lie in the range 0–1. So this formula also standardizes data set. The *daisy* function in the *cluster* package in R programming language used for this task. After that, a clustering algorithm can be selected.

2.1.2 Clustering Algorithm

There is no clustering technique that is universally applicable in covering the variety of structures present in multidimensional data sets and there are no unique objective 'true' or 'best' clusters in a data set [10, 13].

We used hierarchical clustering and Ward's algorithm. Because hierarchical clustering does not need specifying the number of clusters (k) in advance and the user decides for which level represents a good clustering result, this technique is used for segmenting the retail stores. Also, a dendrogram makes hierarchical clustering popular, providing a highly interpretable description of the clustering result [14] (Fig. 4).

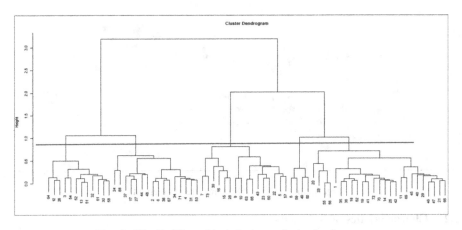

Fig. 4. Ward's Hierarchical cluster analysis: dendrogram

Wards method (1963), the only one among agglomerative clustering algorithm that is based classical sum of error squares criterion, produces groups that minimize within-group dispersion at each binary merging [15]. R programming language has a package, *stats* with *hclust* function for Ward's method.

Below is the dendrogram obtained by Ward's method. When clusters are examined, to have 5 groups for 73 stores seems meaningful, each cluster has different characteristics in terms of the store features or people residing around stores (trade area demographics). Figure 5 shows the map of Istanbul and cities around. Different markers are used for representing each cluster. Markers indicate the location of 73 stores that we are analyzing. In Fig. 6, the stores in Istanbul are shown. As seen in

Fig. 6 obviously, although some stores are very close to each other, they were segmented in different groups (Fig. 4).

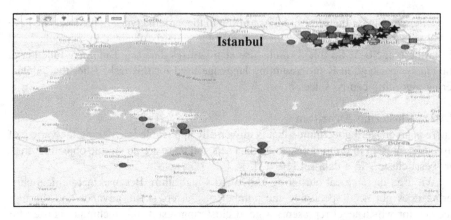

Fig. 5. Segmented 73 stores, each segment indicated with different markers.

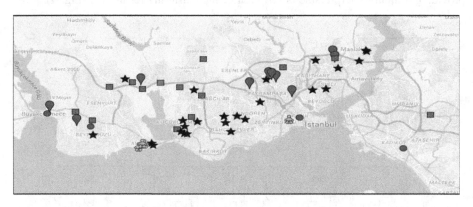

Fig. 6. In Istanbul, 58 stores are segmented.

2.2 Characteristics of Segments

We tried to find the characteristics of each cluster in terms of variables that we have used for clustering. One can see the dominant characteristics of each segment, as shown in the Table 2 below. The population of Istanbul is about 20 million but the land area it distributed is narrow. That is why demographics of the people are different, even for the districts that are neighbor to each other.

3 Analyses for Target Marketing

Frequent itemsets and association rules analyses support supermarket companies for some important decisions such as which items to put on sale, how to design flyers, and how to place items on shelves in order to maximize profit. The Apriori algorithm is the

Table 2. Characteristics of segments, marker

Cluster	Marker	#of Stores	Characteristics
1	●	12	Small sized stores, high rate of 0-19 aged people, low education level
2	●	16	Small sized stores, high rate of 40-59 and 60+ aged people
3	■	16	Large sized stores, high rental rates, car parking, low rate of 60+ aged
4	♔	4	University area, high rate of single people, high competition level
5	★	25	Middle sized stores, bus service, high rental rates, low education level

first and most popular algorithm for mining frequent itemsets and association rules over transactional databases which satisfy minimum support and minimum confidence levels specified by users in advance [17].

The Apriori algorithm first counts the item occurrences for the entire dataset for exploring frequent 1-itemsets, given a support threshold. It works by the principle of "A subset of a frequent itemset must also be a frequent itemset" i.e., if {XY} is a frequent itemset, both {X} and {Y} should be a frequent itemset.

Then, since Apriori uses an iterative approach known as level-wise search; as stated it starts for 1-itemsets (k = 1) and k itemsets is used for discovering k + 1 itemsets till no more frequent k-itemsets can be found. When frequent itemsets are obtained, it is easy to generate interesting association rules with confidence levels larger or equal to a minimum confidence specified by the user.

Association Rule $X \Rightarrow Y$ where $X, Y \subseteq I$ and $X \cap Y = \emptyset$, states when X occurs, Y occurs with a certain probability, X is called antecedent item while Y is called consequent item, the rule means X implies Y.

Support of a rule $X \Rightarrow Y$: Percentage of transactions that contain $X \cup Y$, to the total number of transactions; $supp = Pr(X \cup Y)$.

Confidence of a rule $X \Rightarrow Y$: Percentage of the number of transactions that contain $X \cup Y$ to the number of transactions that have X; $conf = Pr(Y|X) = supp(X \cup Y)/supp(X)$.

In the R programming language, the *apriori* function in *arules* package is used for association rules analyses.

The supermarket chain company that provided the data for our research, limited the transaction data of each store to only for 1 day because of the privacy issues. For the transaction data we have, first we have found the "top 20 frequent 1-item, 2-items and 3-items" for the entire data set (73 stores) and for each 5 clusters. Support measure is arranged each time to have top 20 item/itemsets.

As seen in Tables 3 and 4, clusters have different frequent items and rules too as they have many common items and rules. Details can be found in the Attachment in Appendix (Tables 5 and 6).

Table 3. Unique, frequent 1 items, 2 itemsets and 3 itemsets for each segment

Cluster	# of Transactions	Unique 1 items	Unique 2 itemsets	Unique 3 itemsets
1	1339	margarine	–	cheese-soda-vegetable
				cheese-chocolate-vegetable
				cheese-fruit-soda
2	1554	beef	baking pr.-choco.	baking pr.-chocolate-milk
			Juice-soda	baking pr-cheese-milk
			soda-teaCoffee	baking pr-milk-teacoffee
			baking-cheese	chocolate-soda-teacoffee
			beef-veget	biscuit-chocolate-teacoffee
3	3284	cleaning product dessert	bread-veget	bread-soda-vegetable
			bread-fruit	bread-cheese-vegetable
			bread-soda	baking pr-fruit-vegetable
			bread-milk	
			bread-choc.	
			bread-cheese	
4	317	dressing	teacoffee-veget.	chocolate-milk-vegetable
		napkin	eggs-veget.	dressing-fruit-vegetable
		sausage		cheese-eggs-vegetable
5	3372	–	legume-veget.	milk-soda-vegetable
			chicken-veget.	chocolate-milk-soda

The retail company should consider these frequent unique itemsets and rules of each cluster for advertisement purposes to be able to have successful target marketing. Because of having many frequent itemsets and rules, to better understand the differences of clusters, we should analyze the items sold in terms of their prices for each cluster.

In Tables 5 and 6 in the attachment, the column called Global indicates results for the entire data set; common column is for items/rules which are common for all clusters. The columns for cluster 1, cluster 2... cluster 5, indicate frequent items and rules that are "unique" to that cluster only.

Table 4. Unique, association rules for each segment

Cluster	Unique 2 item rule	Unique 3 item rules
1	{dessert} => {milk}	{fruit,milk} => {vegetable}
	{olives} => {vegetable}	{milk,vegetable} => {fruit}
	{dressing} => {soda	{cheese,fruit} => {vegetable}
2	{stationary} => {chocolate}	{cheese,milk} => {chocolate}
	{cereals} => {milk}	{cheese,soda} => {vegetable}
	{fruitymilk} => {milk}	
	{fruitymilk} => {choc}	
	{pickle} => {c.veget}	
3	{yogurt} => {fruit}	{legume, vegetable} => {fruit}
	{yogurt} => {veget}	{biscuit,milk} => {chocolate}
	{legume} => {veget}	{cheese,milk} => {baking product}
	{baking} => {veget}	{baking product,chocolate} => {milk}
	{cheese} => {veget} {nuts} => {fruit}	
	{veget} => {fruit} {bread} => {veget}	
4	{chicken} => {veget}	{chocolate,milk} => {vegetable}
	{biscuit} => {veget}	{chocolate,vegetable} => {milk}
	{c.veget} => {veget}	{milk, vegetable} => {chocolate}
	{eggs} => {cheese}	{chocolate,vegetable} => {fruit}
5	{butter} => {cheese}	{biscuit,soda} => {chocolate}

4   Conclusion and Future Work

As seen on Istanbul map, in Fig. 6, although some stores are close to each other, they may have different characteristics, so they may be segmented in different groups. Additionally as a result, different type of items might be sold frequently in different stores while similar ones might be sold in similar stores which should lead the retail company to focus target marketing rather than mass marketing approach.

In current work, the items in the transaction data were coded as following: All types of milks, breads, juices, sodas, chocolates, fruits vegetables etc. are coded as milk, bread, juice, soda, chocolate, fruit, vegetable, etc.

To be able to better find the differences between clusters in terms of items sold, we suggest recoding the items according to their prices such as, cheap bread – expensive bread, cheap chocolate – expensive chocolate, etc. Since association rules are changing in time; habits of customers and available products are changing and also clusters are changing because of new residents, new stores and competitor stores etc., we should analyze our models in time.

Appendix

Table 5. Frequent item - 2 itemsets - 3itemsets for the entire data and each cluster

	Frequent item		Frequent 2 itemsets		Frequent 3 itemsets
Global	baking pr.	fruit	fruit-veget	nuts-veget	fruit-soda-veget
	biscuit	juice	cheese-	nuts-soda	cheese-fruit-veget
	bread	laundry pr.	veget	cheese-fruit	fruit-milk-veget
	chicken	yogurt	bread-veget	cheese-milk	fruit-nuts-veget
	teacoffee	milk	choc-soda	cheese-choc	bread-fruit-veget
	vegetable	nuts	milk-veget	fruit-soda	chocolate-fruit-veget
	legume	oil	soda-veget	fruit-nuts	fruit-vegetable-yogurt
	cheese	pasta	bread-fruit	milk-soda	fruit-nuts-soda
	choc	soda	choc-veget	fruit-milk	fruit-teaCoffee-veget
	eggs	sugar	choc-fruit	choc-milk	nuts-soda-veget
			biscuit-choc		
Common	baking pr.	milk	fruit-vegetable		cheese-fruit veget
	biscuit	nuts	soda-vegetable		choc-fruit-veget
	chicken	pasta	cheese -vegetable		fruit-nuts-veget
	cheese	soda	fruit-soda		fruit-milk-veget
	chocolate	teacoffee	chocolate-soda		fruit-soda-veget
	eggs	vegetable	chocolate-fruit		
	fruit	yogurt	chocolate-milk		
	juice		fruit-nuts		
			fruit-milk		
			biscuit-chocolate		
Cluster 1	margarine		-		cheese-soda-veget
					cheese-choc-veget
					cheese-fruit-soda
Cluster 2	beefLamb		baking pr.-choc		baking-choc-milk
			Juice-soda		baking-cheese-milk
			soda-teaCoffee		baking-milk-teacoffee
			baking-cheese		choc-soda-teacoffee
			beefLamb-veget.		biscuit-choco-teacoffee
Cluster 3	cleaning pr.		bread-veget		bread-soda-veget
	dessert		bread-fruit		bread-cheese-veget
			bread-soda		baking-fruit-veget
			bread-milk		
			bread-choc.		
			bread-cheese		
Cluster 4	dressing		teacoffee-veget.		chocolate-milk-veget
	napkin		eggs-veget		dressing-fruit-veget
	sausage				cheese-eggs-veget
Cluster 5	-		legume-veget.		milk-soda-veget
			chicken-veget.		choc-milk-soda

Table 6. Association rules with support and confidence measures, for the entire transaction data and for each cluster

	2 item rules, Supp, Conf	3 item rules, Supp, Conf
Global	{cake}=>{biscuit} 0.03, 0.54 {cake}=>{choc.} 0.034, 0.622 {saltine}=>{biscuit} 0.034, 0.567 {saltine}=> {choc.} 0.035, 0.582 {olives} => {cheese}0.042, 0.627 {sausage}=>{cheese} 0.499, 0.578 {chips}=>{soda} 0.049, 0.545 {biscuit} =>{choc.} 0.083, 0.539 {fruit}=>{vegetable}0.147, 0.524	{bread,fruit}=>{veget} 0.051, 0.68 {bread,veget}=>{fruit} 0.051, 0.571 {fruit, nuts} => {veget} 0.052, 0.63 {nuts,veget}=>{fruit} 0.052, 0.69 {cheese,fruit}=>{veget} 0.053, 0.712 {fruit,soda}=>{veget} 0.058, 0.613 {cheese, vegetable}=> {fruit} 0.053, 0.59 {fruit,milk} =>{veget} 0.053, 0.63 {milk,vegetable}=>{fruit} 0.053, 0.63 {choc,veget}=>{fruit} 0.05, 0.614 {choc,fruit}=>{veget} 0.05, 0.59 {soda,vegetable} =>{fruit} 0.058, 0.579
Common	-	{fruit,milk} => {vegetable} {milk,vegetable} => {fruit} {cheese,fruit} => {vegetable}
Cluster 1	{dessert}=>{milk} 0.047, 0.512 {olives}=>{veget} 0.037, 0.51 {dressing}=>{soda} 0.032, 0.5	{cheese,milk} => {chocolate} 0.036, 0.5 {cheese,soda} => {vegetable} 0.037, 0.55
Cluster 2	{stationary}=>{choc} 0.005, 0.615 {cereals}=>{milk} 0.01, 0.629 {fruitymilk}=>{milk} 0.007, 0.5 {fruitymilk}=>{choc} 0.008, 0.54 {pickle}=>{can veget} 0.009, 0.5	{legume, veget.} => {fruit} 0.022, 0.614 {biscuit,milk} => {choc.} 0.023, 0.55 {cheese,milk} => {baking prod.} 0.022, 0.53 {baking prod.,choc.} => {milk} 0.025, 0.5
Cluster 3	{yogurt}=>{fruit} 0.079, 0.516 {yogurt}=>{veget} 0.087, 0.566 {legume}=>{veget} 0.082, 0.529 {baking}=>{veget} 0.093, 0.5 {cheese}=>{veget} 0.12, 0.539 {nuts}=>{fruit} 0.132, 0.54 {veget}=>{fruit} 0.214, 0.578 {bread}=>{veget} 0.169, 0.5	
Cluster 4	{chicken}=>{veget} 0.063, 0.512 {biscuit}=>{veget} 0.05, 0.57 {can.veget}=>{veget} 0.056, 0.69 {eggs}=>{cheese} 0.06, 0.52	chocolate,milk} => {vegetable} 0.044, 0.63 {chocolate,vegetable} => {milk} 0.044, 0.56 {milk, vegetable} => {chocolate} 0.044, 0.53 {chocolate,vegetable} => {fruit} 0.047, 0.6
Cluster 5	{butter}=> {cheese} 0.02, 0.5	{biscuit,soda} => {chocolate} 0.034, 0.62

References

1. Chen, I.J., Popovich, K.: Understanding CRM, people, process and technology. J. Bus. Process Manage. **9**(5), 672–688 (2003)
2. Ling, R., Yen, D.C.: Customer relationship management: an analysis framework and implementation strategies. J. Comput. Inf. Syst. **41**, 82–97 (2001)
3. Bermingham, P., Hernandez, T., Clarke, I.: Network planning and retail store segmentation, a spatial clustering approach. Int. J. Appl. Geospatial Res. **4**(1), 67–79 (2013)
4. Smith, W.: Product differentiation and market segmentation as alternative marketing strategies. J. Mark. **21**, 3–8 (1956)
5. Segal, M.N., Giacobbe, R.W.: Market segmentation and competitive analysis for supermarket retailing. Int. J. Retail Distrib. Manage. **22**, 38–48 (1994)
6. Berson, A., Smith, S., Thearling, K.: Building Data Mining Applications for CRM. McGraw-Hill, New York (1999)
7. Kolyshkina, I., Nankani, E.: Retail analytics in the context of segmentation, targeting, optimisation of the operations of convenience store franchises. In: Anzmac 2010
8. Mendes, A., Cardoso, M.: Clustering supermarkets: the role of experts. J. Retail. Consum. Serv. **13**, 231–247 (2006)
9. Kumar, V., Karande, K.: The effect of retail store environment on retailer performance. J. Bus. Res. **49**(2), 167–181 (2000)
10. Jain, A.K., Murthy, M.N., Flynn, P.J.: Data clustering: a review. ACM Comput. Surv. **31**(3), 264–323 (1999)
11. Gower, J.C.: A general coefficient of similarity and some of its properties. J. Biometrics **27** (4), 857–871 (1971)
12. Kaufman, L., Rousseeuw, P.J.: Finding Groups in Data: An Introduction to Cluster Analysis. Wiley, New Jersey (1990)
13. Hennig, C., Liao, T.F.: How to find an appropriate clustering for mixed type variables with application to socio-economic stratification. J. Roy. Stat. Soc. Appl. Stat. Part 3 **62**, 309–369 (2013)
14. Hastie, T., Tibshirani, R., Friedman, J.: The Elements of Statistical Learning. Springer, New York (2009)
15. Murtagh, F., Legendre, P.: Ward's hierarchical agglomerative clustering method: which algorithms implement ward's criterion? J. Classif. **31**, 274–295 (2014)
16. Agrawal, R., Imielinski, T., Swami, A.: Mining association rules between sets of items in large databases. In: SIGMOD 1993 Proceedings of the 1993 ACM SIGMOD International Conference on Management of Data, pp. 207–216 (1993)
17. Turkish Statistical Institute. www.turkstat.gov.tr
18. Real Estate Index. www.hurriyetemlak.com
19. www.milliyetemlak.com

Data Mining in Medicine
and System Biology

Searching for Biomarkers Indicating a Development of Insulin Dependent Diabetes Mellitus

Rainer Schmidt[(⊠)]

Institute for Biostatistics and Informatics in Medicine and Aging Research,
University of Rostock, Rostock, Germany
`rainer.schmidt@uni-rostock.de`

Abstract. The aim of our project is to elucidate the mechanisms of the modulating anti CD4 antibody RIB5/2 on prevention of autoimmune destruction of beta cells in the insulin dependent diabetes mellitus rat model. We especially wish to calculate relative risk coefficients for development of overt diabetes at different time points of life. So far, we have applied decision trees and other classification algorithms like random forest and support vector machines on rather small data sets. Furthermore, feature selection was used to support the genes and biomarkers generated by decision trees.

Keywords: Machine learning · Classification algorithms · Bioinformatics · Insulin dependent diabetes mellitus

1 Introduction

Diabetes has been a big challenge in the field of medical informatics for many years. In this paper, we simply say "diabetes" instead of Type 1 diabetes, which is the exact name of the disease of our research. Various programs have been developed concerning diagnosis and treatment of diabetes. The idea of our project deals with an earlier stage of the disease, namely to find good biomarkers that indicate a future development of diabetes. Furthermore, we consider different time points to find out whether the biomarkers may change in time.

From the biological point there are two major challenges for prediction and diagnosis of diabetes. First, though the analysis of various beta cell auto antibodies and beta cell specific T cells allows a good risk assessment for the progression of autoimmunity, biomarkers related to mechanisms of T cell mediated beta cell destruction and induction of self-tolerance are missing. Second, intervention strategies to block beta cell autoimmunity are not fully understood.

Our data are not from human but from rats that were monitored in our laboratory. The insulin dependent diabetes mellitus (IDDM) rat is an animal model of spontaneous autoimmune diabetes which is characterized by a fulminant T cell mediated beta cell destruction leading to a full diabetic syndrome in 60 % of the animals around day 60. The narrow time range of islet infiltration between day 40 and day 50 makes this model a valuable tool to study strategies and mechanisms for induction of immune tolerance.

P. Perner (Ed.): ICDM 2015, LNAI 9165, pp. 47–55, 2015.
DOI: 10.1007/978-3-319-20910-4_4

Induction of immune tolerance is a promising approach to halt autoimmunity in diabetes. Anti CD3 antibodies and vaccination with modified beta cell antigens such as insulin, GAD65, and hsp60 could block autoimmunity and induce self-tolerance in animal models of autoimmune diabetes [1].

These strategies, however, still show limitations that hamper translation into routine clinical use. First, the mechanisms of T cell modulation are still unclear in particular for transition from temporary immune suppression to induction of permanent self-tolerance. Second, despite development of humanized and aglykosylated anti CD3 antibodies the side effects remain severe and raise ethical concerns for treatment of young diabetes patients.

In a first experiment just twelve rats were monitored. For a second just eight and for a third experiment 37 new rats could be bred (but the number decreases in time, because some of the animals died during the experiment). Since different measurement facilities were used, the data of the three experiments could not be merged. In all experiments the rats were monitored for gene expression data in blood immune cells for functional gene clusters on the days 30, 35, 40, 45, 50, 55, 60, 65, 70, 80, and 90 of their life. However, just the days between 45 and 60 are assumed to be important for the prediction whether a rat will develop diabetes.

We used the WEKA environment [2] and applied decision trees and other classification algorithms like random forest and support vector machines on rather small data sets. Furthermore, feature selection was used to support the genes and biomarkers generated by the decision trees.

2 Biological Background

Diabetes is an autoimmune disease in which beta cells are exclusively destroyed by the interaction of antigen presenting cells, T cells, and environmental triggers such as nutrients and viral infection [3, 4]. The destruction of beta cells in diabetes is a complex process comprising a network between beta cells, antigen presenting cells, auto aggressive T cells, and environmental triggers. Beta cells that are under assault are not passive bystanders, but actively participate in their own destruction process [5, 6]. Overall, many of the cytokine- and virus-induced effects involved in inhibition of beta cell function and survival are regulated at the transcriptional and posttranscriptional/translational level [7]. T-cells modulate the autoimmune process and auto reactive T-cells can transfer diseases [8]. Thus, immune intervention during the prodromal phase or at the onset of overt diabetes will affect the balance between auto reactive and regulatory T cells. Currently it is possible to identify ß-cell-specific auto reactive T-cells using standard in vitro proliferation and tetramer assays, but these cell types could also be detected in healthy individuals [9]. Although the analysis of auto antibodies allows an assessment of risk for diabetes, it is still impossible to draw conclusions about T cell function in the local lymphatic compartment of the pancreas. Notably, there is an extensive knowledge upon activation of T cells and upon induction of self-tolerance on the molecular level of gene expression biomarkers. We hypothesize that biomarkers must be analyzed in a dynamic manner because they shall have specific predictive values for development of autoimmunity at different stages of autoimmunity.

The analysis of gene expression patterns might help to distinguish between T1DM affected subjects and healthy animals at an early stage. In a first experiment, we could demonstrate that analysis of selected genes of T cell differentiation, T cell function, and cytokine expression in whole blood cells at an early prediabetic stage (after 45 days of live), the RT6 T cell proliferation gene was most decisive for diabetes onset in the IDDM rat followed by selectin and neuropilin at the stage of islet infiltration (after 50 days), and IL4 during progression of beta cell destruction (after 55 days).

3 Data

Several biological experiments were performed and statistically evaluated. In one of them, for example, it could be shown that the treatment of prediabetic IDDM rats with antibody RIB 5/2 significantly reduces diabetes incidence. This indicates that treatment at a prediabetic status should be successful and supports our intention to find good biomarkers to discover prediabetic patients.

For our first experiments, data from just twelve rats were available. They were monitored for gene expression data in blood immune cells for functional gene clusters on the days 30, 35, 40, 45, 50, 55, 60, 65, 70, 80, and 90 of their life. However, just the days between 45 and 60 are assumed to be important of the prediction whether a rat will develop diabetes or not. Six of the twelve rats developed diabetes, three did not, and another three rats (background strain) were diabetes resistant because of the way they had been bred and because of specific treatment. Unfortunately, due to problems of the measurement facilities the data quality is rather poor. Many data are missing and some are obviously incorrect, especially for the early and the late measurement time points. However, as mentioned above, the most important measurement time points are in the middle. So, for some measurement time points, data from just eleven of the twelve rats were used.

For our second experiment just eight rats and for a third experiment more rats could be bred, which were monitored in the same way as in the first experiment. Since some of them died during the experiment, their number decreases in time: on day 45 there were 37 rats, on day 50 remained 22 and on the days 55 and 60 there were just 16 rats. Since different measurement facilities were used, the data of the three experiments could not be merged.

4 Experimental Results

In the experiments, data of the following measurement time points were used: 45, 50, 55, and 60 days of life. The attributes are eighteen preselected genes and biomarkers. In our first experiment, the class labels are "diabetes", "no diabetes", and "background strain".

4.1   First Experiment

Since we wanted to get attributes that are most decisive for the classification, we applied decision trees, which do not just provide the most decisive attributes but also their splitting point values.

4.1.1   Decision Trees

The C4.5 decision tree algorithm, which was originally developed by Ross Quinlan [10], was applied in form of its J48 implementation in the WEKA environment [2]. Later on, we also applied other classification algorithms that are provided in WEKA, like "random forest", for example.

The tree for day 45 is depicted in Fig. 1. It states that if the expression value of RT6 is between 0.0041 and 0.011, there is a high risk to develop diabetes. Actually the gene Rt6 is already assumed to be responsible for the correct thymic development of T-cells. Since RT6 is decisive at day 45, this indicates that Rt6 may rather at the beginning decide whether autoimmunity will probably develop.

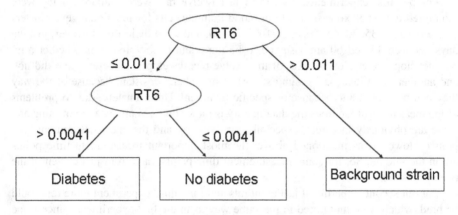

Fig. 1. Decision tree for day 45.

The trees for day 55 and for day 60 are depicted in Figs. 2 and 3. In both trees IL4 is decisive between diabetes and no-diabetes. IL4 is a T cell stimulating cytokine. So, it seems that during the progression of beta cell destruction IL4 is crucial for the progression of beta cell infiltration.

For the three days above, the biochemists involved in our project are quite happy we the results given by the generated decision trees. However, for day 50 this is not really the case. The tree for day 50 is depicted in Fig. 4 and states the following. If the gene expression value of selectin is bigger than 2.14 a rat probably belongs to the background strain, otherwise if the gene expression value of neuropilin is bigger than 0.63 a rat probably does not develop diabetes, otherwise it probably develops diabetes. So, the decisive gene between diabetes and no diabetes is neuropilin.

We did not only apply decision trees but other classification algorithms (see Sect. 4.1.2) and feature selection measures (see Sect. 4.1.3). When applying the feature selection measure InfoGain L-Selectin is on top (for splitting away the background

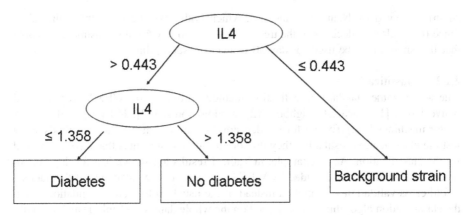

Fig. 2. Decision tree for day 55

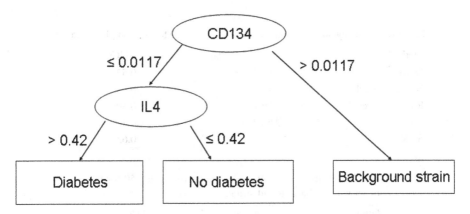

Fig. 3. Decision tree for day 60

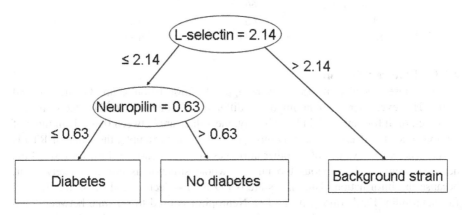

Fig. 4. Decision tree for day 50.

strain), followed by Neuropilin and IL4, which both have the same InfoGain value. Since IL4 is already decisive in the trees of day-55 and day-60, it seems more probable that IL4 should also be used for day-50 instead of Neuropilin.

4.1.2   Classification

Afterwards some standard classification methods provided by WEKA were applied (naïve bayes [11], nearest neighbor [12], random forest [13], J48 [10], and support vector machines [14]). Except for the decision tree algorithm J48 these methods show just the classification results but they do not provide the attributes that have been used for the classification. As an example, in Table 1 results are shown just for day 50. An inner cross validation is provided by WEKA. Because of the small size of the data set 3-fold cross validation was applied instead of the usual 10-fold cross validation. First, the classification algorithms were applied on the whole data sets (Table 1) and secondly on the data sets without background strain (Table 2). However, the differences are very small.

Table 1. Accuracy and area under the curve for day 50 for the complete data set.

Method	Accuracy (%)	AUC
Naïve bayes	58.3	0.52
Nearest neighbor	75	0.75
Random forest	66.7	0.80
J48	66.7	0.76
SVM	58.3	0.65

Table 2. Accuracy and area under the curve for day 50 for the data set without background strain.

Method	Accuracy (%)	AUC
Naïve bayes	55.6	0.42
Nearest neighbor	77.8	0.67
Random forest	66.7	0.53
J48	66.7	0.61
SVM	55.6	0.50

4.1.3   Feature Selection

Three feature selection measures were applied: Information Gain, GainRatio, and Relief. However, there were no important differences in their results. The decision trees are based on Information Gain [2]. Usually, the values are between 0 and 1. In three of the four trees the decision was obvious. For day 45, for example, the value of RT6 is 0.811, whereas the values of all other attributes are 0 or nearly 0. Just, for day 50 the decision seems to be obvious too but the whole situation is not completely clear, because the Information Gain values are 0.959 for L-selection and 0.593 for IL4 and for Neuropilin. Furthermore, in the tree Neuropilin is used to separate between "diabetes" and "no diabetes". So, the background strain was excluded and Information

Gain was used just to classify "diabetes" and "no diabetes", with the result that neuropilin was the first choice.

4.2 Second Experiment

For our second experiment just eight new animals could be bred. Unfortunately, they belong to just two classes, because there is no background strain. Six rats developed diabetes, the other two ones did not. With such a data set the application of decision trees does not seem to be reasonable, because there should be a big chance that an attribute can be found that has the decisive power to split away the two rats from the remaining six diabetes rats.

Actually, the situation is even worse. Most of the preselected attributes can be used to distinguish between the two classes extremely well. Since the decisive attributes we found in the first experiment (RT6, IL4, and Neuropilin) are among them, this new data set supports the findings of the first experiment. However, many other attributes can also be used to separate between the two classes.

4.3 Third Experiment

For our third experiment more rats could be bred, but some of them died during the experiment. So, their number decreases in time: without the background strain on day 45 there were 37 rats, on day 50 remained 22, and on the days 55 and 60 there were just 16 rats. Additionally, there were many background strain rats, e.g. 20 for day-40. These background strain rats could very well split away from the other ones. Since we are interested in genes that are decisive between diabetes and no diabetes, we excluded the background strain from our experiment.

First, we applied the same classification algorithms in the same way as in the other two experiments. Unfortunately, except for day 40 the results (Table 3) are rather poor, even poorer than in the first experiment.

Table 3. Accuracies for days 45, 50, 55, and 60 (without background strain).

Method	Day 45	Day 50	Day 55	Day 60
Naïve bayes	68.2	56.2	66.7	62.5
Nearest neighbor	77.3	56.2	46.7	43.7
Random forest	68.2	68.7	46.7	50
J48	68.2	62.5	53.3	56.2
SVM	81.2	56.2	53.3	62.5

Next we generated decision trees. Since there were more rats the trees were bigger, with some branches. However, when we showed them to our experts, they were not happy with them. They could really explain them and doubted that they contained any valuable results. So, we considered the feature selection measure "information gain", on which the trees are based. Here, we found very low values (about 0.10 on a scale between 0 and 1.0). The decisive attributes we found in the first experiment (RT6, IL4, and Neuropilin) are not on top, but very near the top.

Summarizing the third experiment, we can say: though much more animals could be used, the results are very unsatisfactory.

5 Discussion

The application of Machine Learning (classification) methods has become popular in bioinformatics. This is already reflected at earlier ICDM conferences (e.g. [15, 16]). In our application, the analysis of gene expression patterns might help to distinguish between diabetes affected subjects and healthy animals at an early stage. In a first experiment, we could demonstrate that analysis of selected genes of T cell differentiation, T cell function, and cytokine expression in whole blood cells at an early pre-diabetic stage (after 45 days of live), the RT6 T cell proliferation gene was most decisive for diabetes onset in the IDDM rat model followed by Selectin and Neuropilin at the stage of islet infiltration (after 50 days), and IL4 during progression of beta cell destruction (after 55 days).

In the first experiment, the data set is very small and, probably because of poor data quality, the cross-validated classification results are not really significant (see Tables 1 and 2). Nevertheless, the generated decision trees perform well and nearly all of them can be very well explained by the biochemical experts.

In the second experiment, just eight rats could be breed and the two healthy ones differed from the six diabetes ones in most of the preselected genes. So, the decisive genes found in the first experiment could be somehow supported, because here they can be decisive again but other genes can also be used.

In the third experiment much more animals could be used but the results are very unsatisfactory and do provide probable biomarkers.

References

1. Ludvigsson, J., Faresjo, M., Hjorth, M., et al.: GAD treatment and insulin secretion in recent-onset type 1 diabetes. N. Engl. J. Med. **359**, 1909–1920 (2008)
2. Hall, M., et al.: The WEKA data mining software: an update. SIGKDD Explor. **11**(1), 10–18 (2009)
3. Akerblom, H.K., Vaarala, O., Hyoty, H., Ilonen, J., Knip, M.: Environmental factors in the etiology of type 1 diabetes. Am. J. Med. Genet. **115**, 18–29 (2002)
4. Jun, H.S., Yoon, J.W.: A new look at viruses in type 1 diabetes. Diabetes Metab. Res. Rev. **19**, 8–31 (2003)
5. D'Hertog, W., Overbergh, L., Lage, K., et al.: Proteomics analysis of cytokine-induced dysfunction and death in insulin-producing INS-1E cells: new insights into the pathways involved. Mol. Cell. Proteomics **6**(21), 80–99 (2007)
6. Rasschaert, J., Liu, D., Kutlu, B., Cardozo, A.K., Kruhoffer, M., Ørntoft, T.F., Eizirik, D.L.: Global profiling of double stranded RNA- and IFN-gamma-induced genes in rat pancreatic beta cells. Diabetologia **46**, 1641–1657 (2003)

7. Gysemans, C., Callewaert, H., Overbergh, L., Mathieu, C.: Cytokine signalling in the beta-cell: a dual role for IFNgamma. Biochem. Soc. Trans. **36**, 328–333 (2008)
8. Lampeter, E.F., McCann, S.R., Kolb, H.: Transfer of diabetes type 1 by bone-marrow transplantation. Lancet **351**, 568–569 (1998)
9. Schloot, N.C., Roep, B.O., Wegmann, D.R., Yu, L., Wang, T.B., Eisenbarth, G.S.: T-cell reactivity to GAD65 peptide sequences shared with coxsackie virus protein in recent-onset IDDM, post-onset IDDM patients and control subjects. Diabetologia **40**, 332–338 (1997)
10. Quinlan, J.R.: C4.5 Programs for Machine Learning. Morgan Kaufmann, San Mateo (1993)
11. Gan, Z., Chow, T.W., Huang, D.: Effective gene selection method using bayesian discriminant based criterion and genetic algorithms. J. Sign. Process. Syst. **50**, 293–304 (2008)
12. Cost, S., Salzberg, S.: A weighted nearest neighbor algorithm for learning with symbolic features. Mach. Learn. **10**(1), 57–78 (1993)
13. Breiman, L.: Random forest. Mach. Learn. **45**(1), 5–32 (2001)
14. Platt, J.: Advances in Large Margin Classifiers, pp. 61–74. MIT-Press, Cambridge (1999)
15. Bichindaritz, I.: Methods in case-based classification in bioinformatics: lessons learned. In: Perner, P. (ed.) ICDM 2011. LNCS, vol. 6870, pp. 300–313. Springer, Heidelberg (2011)
16. Perner, J., Zotenko, E.: Characterizing cell types through differentially expressed gene clusters using a model-based approach. In: Perner, P. (ed.) ICDM 2011. LNCS, vol. 6870, pp. 106–120. Springer, Heidelberg (2011)

Predictive Modeling for End-of-Life Pain Outcome Using Electronic Health Records

Muhammad K. Lodhi[1], Janet Stifter[1], Yingwei Yao[1], Rashid Ansari[1],
Gail M. Keenan[2], Diana J. Wilkie[2], and Ashfaq A. Khokhar[3(✉)]

[1] University of Illinois at Chicago, Chicago, IL, USA
{mlodhi3,jstift2,yyao,ransari}@uic.edu
[2] University of Florida, Gainesville, FL, USA
{gkeenan,diwilkie}@ufl.edu
[3] Illinois Institute of Technology, Chicago, IL, USA
ashfaq@iit.edu

Abstract. Electronic health record (EHR) systems are being widely used in the healthcare industry nowadays, mostly for monitoring the progress of the patients. EHR data analysis has become a big data problem as data is growing rapidly. Using a nursing EHR system, we built predictive models for determining what factors influence pain in end-of-life (EOL) patients. Utilizing different modeling techniques, we developed coarse-grained and fine-grained models to predict patient pain outcomes. The coarse-grained models help predict the outcome at the end of each hospitalization, whereas fine-grained models help predict the outcome at the end of each shift, thus providing a trajectory of predicted outcomes over the entire hospitalization. These models can help in determining effective treatments for individuals and groups of patients and support standardization of care where appropriate. Using these models may also lower the cost and increase the quality of end-of-life care. Results from these techniques show significantly accurate predictions.

Keywords: Electronic health records (EHR) · Data mining · Predictive modeling · End-of-life (EOL)

1 Introduction

The ability to predict the condition of a patient during hospitalization is crucial to providing adequate and cost effective care. It is heavily influenced by diverse factors including the patient's personal as well as psychological characteristics, and other health problems. Different data mining algorithms have been used to help identify characteristics routinely accompanying select patient conditions.

In recent years, there has been an increasing use of electronic health records (EHR) in the healthcare industry. Historically, in most cases, EHRs are merely used for monitoring the progress of patients [1, 2]. However, according to PubMed [3], since 2005, a plethora of research work has been pursued related to the development of prediction models using EHR data. As EHR systems are quite large in size and contain a variety of historical data, they are ideal candidates to study Big Data issues, including data analytics, storage, retrieval techniques, and decision making tools. In the U.S.,

© Springer International Publishing Switzerland 2015
P. Perner (Ed.): ICDM 2015, LNAI 9165, pp. 56–68, 2015.
DOI: 10.1007/978-3-319-20910-4_5

more than $1.2 trillion is wasted in healthcare annually, out of which $88 billion goes to waste because of ineffective use of technology [4]. Discovering the hidden knowledge within EHR data for improving patient care offers an important approach to reduce these costs by recognizing at-risk patients who may be aided from targeted interventions and disease prevention treatments [5].

One important application of predictive modeling is to correctly identify the characteristics of different health issues by understanding the patient data found in EHR [6]. In addition to early detection of different diseases, predictive modeling can also help to individualize patient care, by differentiating individuals who can be helped from a specific intervention from those that will be adversely affected by the same intervention [7, 8].

Pain is a very common problem experienced by patients, especially at the end of life (EOL) when comfort is paramount to high quality healthcare. Unfortunately, comfort is elusive for many of the dying patients. Research findings over the past two decades show minimal progress in improving pain control for patients at the EOL [9, 10]. A variety of methodological issues, including the patients' vulnerable health status, make it difficult to conduct prospective pain studies among EOL patients [11, 12]. It is, however, possible that EHR data could provide insights about ways to improve pain outcomes among the dying.

In this paper, we focus on the analysis of nursing care data within EHR systems. Evaluating nursing data in the EHR can help guide in more effective management of patients and thus help produce cost savings and better patient outcomes. Unfortunately, most of the data that is currently entered by the nurses is not analyzable due to the absence of comparability in data collection practices. Since nurses are the main front line providers of care, understanding their care and the impact of it is crucial to overall healthcare.

There are a number of examples in literature that have used data mining for decision making models [13]. However, in those papers, numerous problems were reported, mostly because the storage of data was not in a standardized format. Hsia and Lin [14] identified the relationship between different nursing practices and related function. Using mining of correlations present among nursing diagnosis, nursing outcomes and nursing interventions, care plan recommendations were proposed in [15].

In this paper, we examine pain management of EOL patients. This group of patients is of significant importance as it takes a disproportionate share of healthcare expenses. 30 % of the Medicare expenses go to just 5 % of all the beneficiaries who die every year, mostly because of inappropriate and often unwanted treatments provided to the patients [16]. These costs are expected to rise substantially as the baby boomer generation ages [17], unless treatment is provided more efficiently. It is therefore imperative to make use of the knowledge hidden in EHR systems to deliver cost effective and high quality pain care to patients at the EOL. In this paper, we explore the design of data models to determine predictors that influence pain outcome in EOL patients. The goal is to determine the associations between a variety of nursing care interventions or patient characteristics and improved outcomes for EOL patients. These models can then be used to identify best practices and then to predict likely outcomes in real world environments [18].

2 Data Description

The data used in our analysis were collected from nine medical-surgical units located in four unique Midwestern U.S. hospitals during a three-year study (2005–2008) using the Hands-On Automated Nursing Data System (HANDS) [18]. The HANDS is an EHR system that is designed specifically to record nursing care provided to the patients using standard terms. In HANDS, diagnoses are coded based on nursing diagnosis from North American Nursing Diagnosis Association – International (NANDA-I), which is a professional organization of nurses offering standardized nursing terminology for different nursing diagnoses [19]. Similarly, different nurse-oriented outcomes related to different NANDA-Is are coded based on the Nursing Outcome Classification (NOC) [20], and nursing interventions, related to different NOCs are coded using the Nursing Interventions Classifications (NIC) [21].

There were a total of 42,403 episodes (34,927 unique patients) recorded in the three-year span. An episode is defined as continuous patient stay in a single hospital unit and consists of multiple nurse shifts. Each nurse shift corresponds to a single plan of care (POC) that the nurses document at every formal handoff (admission, shift change, discharge). Every POC contains the nursing diagnoses (NANDA-I) for the patient. For every diagnosis, different outcomes (NOCs) are identified with an expected score for discharge (assigned to each unique NOC on the POC to which it was initially entered) and a current rating assigned at handoff. For every outcome, there are different nursing interventions (NICs) used to achieve the expected outcome. The POC also stores demographic information about the patient and nurses providing care to the patient. The Table 1 gives two example episodes, containing multiple POCs that nurses documented for each episode.

Table 1. Example of HANDS database

Episode ID	Shift #	NANDA	NOC	NIC	Patient age	Nurse experience (years)	Current NOC rating	Expected NOC rating
1	1	Acute pain	Pain level	Comfort care	56	3	3	5
1	2	Acute pain	Pain level	Comfort care	56	2	4	5
2	1	Chronic pain	Pain level	Comfort care	90	5	2	3

Out of the 42,403 episodes available in our database, a subset of episodes belonging to the EOL patients was chosen for our experiments. An episode is considered to be an EOL episode when any of its POCs contains one or more of the following indicators:

- Discharge Status = "expired" or "hospice home care" or "discharge to hospice medical facility"
- NOC = "Comfortable Death" or "Dignified Life Closure"
- NIC = "Dying care".

There were a total of 1,453 unique EOL patients, a few with multiple episodes. Only the last episode for patients with multiple episodes was considered; therefore each patient is represented only once in the dataset. Past episodes for patients with multiple episodes were not included as there were only eight such patients.

Of these 1453 patients, 652 patients, whose episodes included multiple POCs, were diagnosed with either Acute or Chronic Pain. The patients with a single POC were excluded because a minimum of two POCs are needed to evaluate a patient's progress. The set of patients was further decreased by including only those patients that had a NOC: Pain Level. The final sample includes 160 unique patient episodes, containing a total of 1,617 POCs.

3 Feature Extraction

In this paper, predictive modeling and analyses have been conducted at two different levels of granularity. The first analysis deals with predictive modeling at the episode level. The second analysis is more fine-grained and is performed at the POC level. The target variable for predictive modeling is "NOC met", and it is defined as follows:

NOC met: A NOC is considered "met" if the final NOC outcome rating is the same or better than the NOC goal or expected rating set by the nurse at the beginning of the episode. Otherwise it is considered "not met". For episode level modeling, it is computed using the NOC rating set at the time of discharge or death. For POC level fine grained modeling it is computed based on NOC rating at the end of each shift.

The aim of the study is to develop predictive models that determine how different variables impact a NOC outcome. The predictor variables extracted from HANDS data and used in our analysis include the following: the initial NOC outcome rating, expected NOC outcome rating, the final NOC outcome rating, patient's age, length of stay (LOS) and nurse experience. In the dataset, initial, expected, and final NOC outcome ratings vary between 1 and 5, with 1 being the worst outcome. Patients' age, LOS, and nurse experience were continuous variables that were discretized for our analysis.

To study clinically meaningful impact of the patients' age, we divided age into four age groups: young (18–49), middle-aged (50–64), old (65–84), and very old (85+) based on theoretical [22] and the data frequency distribution.

LOS for each episode was a derived variable in the database. It was calculated by accumulating the number of hours for all the nurse shifts in the episode. LOS was then divided into three categories: short (up to 48 h), medium (48–119 h) and long stay (120 + h). Currently, in most places, visits less than 48 h are called observation visits. Average LOS of patient in the hospital is usually around 120 h, which in our case is considered as a medium stay. Anything longer than 120 h is therefore considered as a long stay [23].

Average nurse experience was also a derived variable. For each episode, we summed up the experience of all the nurses that provided care and then divided by the total number of different nurses in that episode. For nurse experience, a nurse with at least two years of experience in her current position was considered to be an experienced nurse, and the nurses with less than two years' experience to be inexperienced.

The episode was categorized as care provided by an experienced nurse if more than 50 % of the nurses providing care in that episode had at least 2 years of experience. These categories were based on professional criteria [24].

Along with these patients and nurse characteristics, the NANDA-I diagnoses and NIC interventions that appeared in the POCs were also included as predictive variables. Given the large number of unique NANDA-I diagnoses and NIC interventions, they were clustered together by NANDA and NIC domains, respectively, based upon the respective taxonomic structures of these systems [19, 21]. The NANDA-I terms in our dataset, that were used for this study population fell within eight of the twelve NANDA-I domains (Activity/Rest, Comfort, Coping/Stress Tolerance, Elimination, Health Promotion, Nutrition, Perception, and Safety/Protection). Similarly, the data analyzed in this study included terms from six out of seven NIC domains (Behavioral, Family, Health System, Safety, Physiological: Basic, and Physiological: Complex). A particular NANDA-I or NIC domain was considered to be either present or absent in an episode. The other NANDA-I and NIC domains were excluded due to low frequency counts.

4 Data Modeling

Our aim for conducting these experiments was to construct models that can predict the pain level outcome for EOL patients. The secondary objective was to assess the feasibility of constructing predictive models using data in the HANDS database.

After data extraction and data refinement, multiple models were constructed on the dataset using different prediction tools and their performances were compared. In the first set of experiments, the models were based on decision trees [25], k-nearest neighbors (k-NN) [26], support-vector machines (SVM) [27] and Logistic Regression (LR) [26]. Naïve-Bayes models [28] were not included in the episode-level analysis because our definition of "NOC met" is directly related to Expected NOC outcome rating, and hence the Naïve Bayes conditional independence assumption would be violated [29]. Naïve-Bayes modeling was however performed in experiments on POC level. RapidMiner 6.0 was used to implement all the data modeling techniques discussed in this paper.

5 Experimental Results

The initial analysis was to develop a model that predicts whether the NOC: Pain Level was met based on different patient and nurse characteristics or different diagnoses and interventions. As such, the target variable was a binary variable (met or not met). In the first set of experiments, the final NOC outcome rating was not included as a predictor since the target variable is a deterministic function of the final and expected NOC outcome ratings.

The performance of the models developed in this study was evaluated using 10-fold cross-validation [30]. The models were also evaluated using the f-measure, precision, recall, and specificity. In our case, the positive class was "NOC met".

5.1 Coarse-Grained Episode Level Models

Predicting Expected NOC Outcome Rating. We first present the results for the coarse grain models that predict whether NOC is met or not met at the end of an episode. The performance results based on different measures are shown in Table 2.

Table 2. Model performance for experiment 1: predicting expected NOC outcome rating (met/not met).

Model used	Accuracy (%)	F Measure	Precision	Recall	Specificity
Decision tree (C 4.5)	90.6	91.4	89.9	93.0	87.8
k-NN (k = 5)	74.4	76.8	74.7	79.1	68.9
k-NN (k = 1)	98.1	98.3	97.7	98.8	97.3
SVM	96.9	97.1	96.6	97.7	95.9
LR (evolutionary)	74.4	80.4	68.3	97.7	47.3

The accuracies for all the models are high, with all the models having prediction accuracy of more than 74 %. K-NN (k = 1) had the best accuracy, followed by SVM model. Other measures were generally high in most cases. There were, however, a few exceptions as seen. These initial results indicate that the data collected by the HANDS EHR have predictive capacities.

To determine if there was a significant difference in the results of these different predictive models, Pearson's two proportions z-test [31] is used. In our experiments, if p-value is less than 0.05, we consider the difference to be statistically significant.

Using Pearson's two proportions z-test, the results of k-NN (k = 1) and SVM (p-value: 0.43) and the results of SVM and decision tree (p-value: 0.07) are not statistically different. However, the difference between k-NN (k = 1) and decision tree was seen to be statistically significant (p-value: 0.01). The difference between decision tree and LR/k-NN (k = 5) (p-value < 0.0002) was also very significant. Other comparison results are not given, since k-NN (k = 5) and LR have same accuracy, and since decision tree is better than LR/k-NN (k = 5), we can safely assume that k-NN (k = 1) and SVM will also be significantly better.

Feature Selection for Experiment 1. Feature selection for this experiment was done using Least Absolute Shrinkage and Selection Operation (LASSO) method [32]. The Lasso is a shrinkage and selection method for linear regression. It minimizes the sum of squared errors, with a bound on the sum of the absolute values of the coefficients. Using this method, initial and expected NOC rating were given the highest weight, implying these two attributes were the most important in the predictive models. Age and LOS were also important patients' attributes using LASSO. Nurse experience was given a weight of 0, implying that nurse experience attribute did not play an important role in prediction results. Among NANDA domains, Coping/Stress Tolerance domain was given the highest weight, followed by Perception/Cognition, and Activity domain. Psychological: Basic and Safety were the only NIC domains that were selected as features for prediction. Other NANDA and NIC domains were not selected as attributes as these attributes or features were assigned a weight of zero, marking them as irrelevant features for the predictive models for our experiments.

Predicting Final NOC Outcome Rating. After the initial experiment, another experiment was conducted, also at the episode-level. This time, we tried to determine whether the Final NOC rating can be predicted accurately using the same data. For this analysis, the final NOC outcome rating was set as a target variable (values between 1 and 5). "NOC met" variable was not included in the analysis. The results for the second analysis were also evaluated using 10-fold cross validation. Not all of the performance metrics could be used due to the polynomial target attribute. The results appear in Table 3.

Table 3. Model performance for experiment 2: Predicting the final NOC rating

Model used	Accuracy %
k-NN (with k = 1)	80.0
K-NN (with k = 5)	72.5
Decision tree	85.0
SVM	69.4
LR	68.1

The accuracies were high with k-NN and decision tree models achieving around an 80 % accuracy rate. SVM and LR models produced the worst accuracy results for these experiments. The results of other models again showed the predictive capacity of the data in the HANDS system.

Using Pearson's proportion test again, we discovered that the difference between accuracies of decision tree and k-NN (k = 1) was statistically significant (p-value < 0.0002). Similarly, difference between k-NN (k = 1) and k-NN (k = 5) was statistically significant (p-value: ∼ 0). The difference between k-NN (k = 5) and SVM was not deemed significant (p-value: 0.052), whereas the difference between k-NN (k = 5) and LR was significant statistically (p-value: 0.006). Other pair-wise comparison results are not given since it is safely assumed that if decision tree is better than k-NN (k = 1), decision tree will be significantly better with other results as well.

5.2 Fine-Grained Plan of Care Level Models

After the initial analyses at episode-level, we drilled further down at the care plan level. The aim was to predict the current NOC rating for each nurse shift (i.e. POC level) within each episode based on the previous nurse shift rating and factors in the POC (age, nurse experience, NANDA and NIC domains) related to the nurse shift. LOS was not included in this analysis as the analysis now is being performed for each nursing shift in a patient's plan of care. Along with decision trees, k-NN, SVM, and LR, Naïve-Bayes model was also used. Table 4 shows the accuracies of different models. As seen, SVM, k-NN (k = 5) and LR models were the worst performing models.

The accuracy for remaining models was around 70 % with Naïve-Bayes giving the best accuracy results. After two proportion z-test, it can be stated that accuracy results of decision tree and Naïve-Bayes were not statistically different (p-value: 0.37).

Table 4. Model performance for experiment 3: predicting the NOC rating of current shift

Model used	Accuracy %
k-NN (with k = 1)	68.1
k-NN (with k = 5)	65.9
Decision tree	70.1
Naïve-Bayes	71.5
SVM	66.3
LR	65.2

Similarly, statistically, there was no difference in accuracy results of decision tree and K-NN (k = 1) (p-value: 0.22). K-NN (k = 1) and Naïve-Bayes, when their results are compared using z-test, there is a statistically significant difference (p-value: 0.034). This also implies that the result of Naïve-Bayes is significantly better than k-NN (k = 5), SVM and LR, models. Comparing decision tree results with other low performing models, there is significant difference between decision tree and SVM (p-value: 0.02) and between k-NN (k = 5) and decision tree (p-value: 0.01).

The prediction accuracies of all the models were lower than what we expected. Based on these results, the POCs were then further divided into two groups, based on the timing of the shifts within an episode. The first group consisted of first half of the shifts in an episode whereas the second group consisted of the second half of the shifts in the episode. We did so because we thought that prediction accuracy might be higher in later POCs since the ratings should be getting closer to the expected outcome. As expected, we noted that the accuracies for the second group were higher when compared with the first group. For these experiments, we did not consider k-NN (k = 5), LR and SVM model because of their poorer result in the previous analysis. The accuracy results are shown in a Fig. 1.

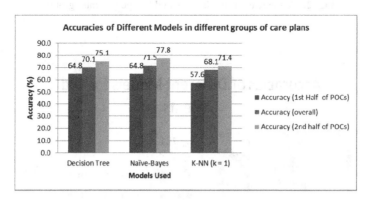

Fig. 1. Accuracies of different models in different groups of care-plans

Using two proportions z-test once again, we tried to determine if the difference between the two groups was statistically significant. For all the models, the difference between the two groups was statistically significant.

The models were then built based on different age groups to see whether different age groups have any impact on the accuracy of the models. POCs of young patients were excluded as there were only 37 POCs for young patients. On the other hand, there were 358 POCs for middle-aged patients, 880 for the old, and 342 for the very old patients. These groups were further sub-divided into two groups each based on shift numbers as done previously. The figures below show the accuracy difference between subgroups for each age group.

As seen from the Figs. 2, 3, and 4, the first half of POCs for all age groups have a lower accuracy compared with the group that contains the second half of the shifts. It was also noted that the accuracy was best for old patients, achieving more than 78 % accuracy in decision tree model, while the accuracies for the middle-aged and very old patients was about the same. It was our understanding that since prediction accuracy is better for later POCs, age groups with higher percentage of longer episodes would likely have better prediction accuracy.

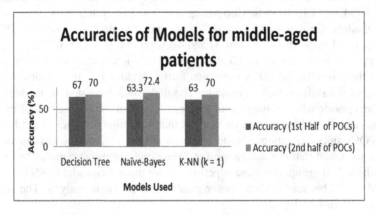

Fig. 2. Accuracies of models for middle-age patients' group

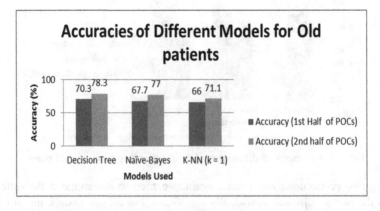

Fig. 3. Accuracies of models for old patients' group

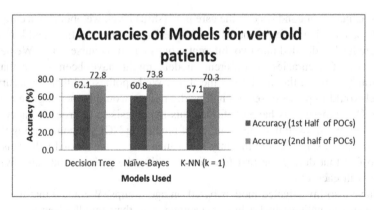

Fig. 4. Accuracies of models for the very old patients' group

When we investigated the age groups and LOS together, we found that the old patients do not have the highest percentage of long episodes, but they do have the highest percentage of long and medium episodes which may have resulted in higher accuracy for later POCs of old patients.

In all the experiments conducted above, Decision trees have consistently performed better. Although in the first experiment, K-NN (k = 1) performed better than decision tree statistically, k-NN (k = 1) has not performed better in the later experiments. Naïve-Bayes performed better in the last experiment, though it was as good as the decision tree model statistically. Also, Naïve-Bayes models cannot be used in our episode-level analysis. Therefore, for our data, we purpose decision tree algorithm to be used for predictive modeling.

6 Discussion and Conclusion

Prediction models are gaining prominence in different fields, including the field of nursing. The adoption of EHRs and the volume and detail of patient information is growing at a very fast pace. As the size of medical data in general, and nursing care data in particular, increases, predictive models over data derived from EHRs will become more prevalent in the nursing field. These predictive models are also important for individual diagnosis and decision making.

In this paper, our objective was to build prediction models that can predict outcomes for the pain level outcome, an important nursing NOC for EOL patients. Our goal was also to test the feasibility of these predictive models using existing nursing data in an EHR system. Different models, that included decision trees, k-NN, Naïve-Bayes, LR, and SVM, were constructed with very high accuracies for our experiment at episode-level. Other performance measures such as precision, recall, f-measure, and specificity were also high for most of the models. In our second experiment on episode-level data, we aimed to predict the final NOC outcome rating. High accuracies were also achieved in this experiment except for SVM and LR models.

In our experiment conducted at the care plan (shift) level, we obtained accuracies of around 70 % by using different models for Naïve-Bayes, decision trees and k-NN. The POCs were further divided into two different groups based on nurse shifts. We believed that the overall accuracies of different models might have been lower than our expectation because of the first half of the shifts. We hypothesized that the accuracy of the model would improve in second half as the ratings would be nearing the expected outcome. Our experiments have confirmed this hypothesis to be true. Using two proportion z-test, we showed that the accuracy of predicting NOC outcome during the first half of the patient stay was significantly lower than that of the second half. This could be due to fact that during the first few nursing shifts the patients and care givers are adjusting with each other.

We have also investigated models based on age groups. We noted that the models consisting of old patients had a higher accuracy rate than middle-aged and very old patients perhaps because old patients had more long and medium length of stays than the other age groups.

Our results from predictive modeling on a large EHR data such as HANDS have shown that these systems have a strong capability to be used as a foundation in nursing decision support framework. Most of our models were built with 70 % or more accuracy for different types of analyses. It is likely that through such approaches, the untapped potential of EHR systems will be fully recognized.

The integration of EHRs with data modeling can augment traditional research methods and improve the prediction accuracy. When EHRs contain consistent data about patients and nurses modeling, can be designed and used for devising efficient nursing patient care. Specifically, the predictive models can help in three different ways. They can help in improving the quality of nursing care, reducing ever-rising healthcare costs, and improving population health in general [33]. Despite the potential limitations in the data, using such models can go a long way in standardizing different treatments and is one significant step towards personalized care.

Disclosure Statement. The HANDS software that was used in this study is now owned and distributed by HealthTeam IQ, LLC. Dr. Gail Keenan is currently the President and CEO of this company and has a current conflict of interest statement of explanation and management plan in place with the University of Illinois at Chicago.

Acknowledgements. This research was made possible by Grant Number 1R01 NR012949 from the National Institutes of Health, National Institute for Nursing Research (NINR). Its contents are solely the responsibility of the authors and do not necessarily represent the official views of the NINR.

References

1. Feliu, J., et al.: Development and Validation of a prognostic nomogram for terminally ill cancer patients. J. Natl. Cancer Inst. **103**(21), 1613–1620 (2011)
2. Gagnon, B., et al.: Montreal prognostic score: estimating survival of patients with non-small cell lung cancer using clinical biomarkers. Br. J. Cancer **109**(8), 2066–2071 (2013)

3. PubMed. http://www.ncbi.nlm.nih.gov/pubmed. Accessed 27 Sept 2014
4. Kavilanz, P.B.: Health Care's Big Money Wasters, 10 August 2009. http://www.money.cnn.com/2009/08/10/news/economy/healthcare_money_wasters/. Accessed 29 June 2014
5. Yeager, D.: Mining untapped data. Record **25**(5), 10 (2013)
6. Bellazi, R., Zupan, B.: Predictive data mining in clinical medicine: current issues and guidelines. Int. J. Med. Inf. **77**(2), 81–97 (2008)
7. Jensen, P.B., Jensen, L.J., Brunak, S.: Mining electronic health records: toward better research applications and clinical care. Nat. Rev. Genet. **13**, 395–405 (2012)
8. Vickers, A.J.: Prediction models in cancer care. Cancer J. Clin. **61**(5), 315–326 (2011)
9. A controlled trial to improve care for seriously ill hospitalized patients. The study to understand prognoses and preferences for outcomes and risks of treatments (SUPPORT). The SUPPORT Principal Investigators. JAMA **274**(20) 1591–1598 (1995)
10. Yao, Y., et al.: Current state of pain care for hospitalized patients at end of life. Am. J. Hosp. Palliat. Care **30**(2), 128–136 (2013)
11. Tilden, V.P., et al.: Sampling challenges in end-of-life research: case-finding for family informants. Nurs. Res. **51**(1), 66–69 (2002)
12. Tilden, V.P., et al.: Sampling challenges in nursing home research. J. Am. Med. Dir. Assoc. **14**(1), 25–28 (2013)
13. Goodwin, L., et al.: Data mining issues and opportunities for building nursing knowledge. J. Biomed. Inform. **36**(4–5), 379–388 (2003)
14. Hsia, T., Lin, L.: A framework for designing nursing knowledge management systems. Interdiscip. J. Inf. Knowl. Manag. **1**, 13–22 (2006)
15. Duan, L., Street, W., Lu, D.: A nursing care plan recommender system using a data mining approach. In: 3rd INFORMS Workshop on Data Mining and Health Informatics, Washington, DC (2008)
16. Zhang, B., et al.: Health care costs in the last week of life: associations with end-of-lofe conversations. Arch. Intern. Med. **169**(5), 480–488 (2009)
17. Smith, T., et al.: A high-volume specialist palliative care unit and team may reduce in-hospital EOL care costs. J. Palliat. Med. **6**(5), 699–705 (2003)
18. Keenan, G., et al.: Maintaining a consistent big picture: meaningful use of a Web-based POC EHR system. Int. J. Nurs. Knowl. **23**(3), 119–133 (2012)
19. NANDA International: Nursing Diagnoses: Definition and Classification 2003–2004. NANDA International, Philadelphia (2003)
20. Moorhead, S., Johnson, M., Maas, M.: Iowa Outcomes Project, Nursing Outcomes Classification (NOC). Mosby, St. Louis (2004)
21. Dochterman, J., Bulecheck, G.: Nursing Interventions Classification (NIC). Mosby, St. Loius (2004)
22. Gronbach, K.W.: The Age Curve: How to Profit from the Coming Demographic Storm. AMACOM Div American Mgmt Assn (2008)
23. Hospital utilization (in non-federal short-stay hospitals): Centers for Disease Control and Prevention (2014)
24. Benner, P.: From novice to expert. Am. J. Nurs. **82**(3), 402–407 (1982)
25. Quinlan, J.: C4.5: Programs for Machine Learning. Morgan Kaufmann Publishers Inc., San Francisco (2003)
26. Aha, D., Kibler, D., Albert, M.: Instance-based learning algorithms. Mach. Learn. **6**(1), 37–66 (1991)
27. Cortes, C., Vapnik, V.: Support-vector networks. Mach. Learn. **20**(3), 273 (1995)
28. Pearl, J.: Bayesian networks. In: Arbib, M. (ed.) The Handbook of Brain Theory and Neural Networks. MIT Pres, Cambridge (1998)

29. Lewis, D.: Naive (Bayes) at forty: the independence assumption in information retrieval. In: Proceedings of 10th European Conference on Machine Learning, pp. 4–15 (1998)
30. Witten, I.H., Frank, E.: Data Mining: Practical Machine Learning Tools and Techniques, 2nd edn. Elsevier, New York (2005)
31. Pearson, K.: On the criterion that a given system of deviations from the probable in the case of a correlated system of variables is such that it can be reasonably supposed to have arisen from random sampling. Philos. Mag. Ser. 5 **50**(302), 157–175 (1900)
32. Tibshirani, R.: Regression shrinkage and selection via the Lasso. J. Roy. Stat. Soc. **58**(1), 267–288 (1996)
33. Desikan, P., et al.: Predictive modeling in healthcare: challenges and opportunities. http:// lifesciences.ieee.org/publications/newsletter/november-2013/439-predictive-modeling-in-healthcare-challenges-and-opportunities. Accessed 27 Sept 2014

Data Mining in Pathway Analysis for Gene Expression

Amani AlAjlan[1]([envelope]) and Ghada Badr[1,2]

[1] College of Computer and Information Sciences, King Saud University, Riyadh,
Kingdom of Saudi Arabia
aalajlan@ksu.edu.sa, badrghada@hotmail.com
[2] IRI - The City of Scientific Research and Technological Applications, Alex, Egypt

Abstract. Single gene analysis looks to a single gene at a time and its
relation to a specific phenotype such as cancer development. However,
pathway analysis simplifies the analysis by focusing on group of genes at
a time that involve in the same biological process. Pathway analysis has
useful applications such as discovering diseases, diseases prevention and
drug development. Different data mining approaches can be applied in
pathway analysis. In this paper, we overview different pathway analysis
techniques in analyzing gene expression and propose a classification for
them. Pathway analysis can be classified into: detecting significant pathways and discovering new pathways. In addition, we summarize different
data mining techniques that are used in pathway analysis.

Keywords: Pathway analysis · Gene expression · Clustering · Classification · Feature selection

1 Introduction

In 2014, statistics from American Cancer Society [6] stated that cancer is the
second cause of death in the United States after the heart diseases. About one
every four deaths are caused by cancer [6]. Recently, pathway analysis has lead
to better cancer diagnosis and treatment.

Pathway is "a collection of genes that serves a particular function and/or
genes that interact with other genes in a known biological process" [29]. In [18]
they defined pathway as "a collection of genes that chemically act together in
particular cellular or physiologic function". In [1] they defined it as "a series of
actions among molecules in a cell that leads to a certain product or a change
in a cell". There are three type of pathways: metabolic, gene regulation and
signaling pathways. Metabolic pathways are series of chemical reactions in cells
[1]. Gene regulation pathways regulate genes to be either active or inhibit [1].
Signaling pathways are series of actions in a cell to move signals from one part
of the cell to another. In biological research, they classify genes in pathways to
improve gene expression analysis [11] and simplify the analysis by looking to few
groups of related genes (pathways) instead of looking to long lists of genes [15].

© Springer International Publishing Switzerland 2015
P. Perner (Ed.): ICDM 2015, LNAI 9165, pp. 69–77, 2015.
DOI: 10.1007/978-3-319-20910-4_6

Pathway analysis concerns with finding out which pathways are responsible for a certain phenotype or which pathways are significant under certain conditions [3]. In addition, pathway analysis is used to explain biological results and as a validation phase in computational research [15]. Khatri et al. [15] pointed out two advantages of using pathway analysis. First, reduce the complexity of analysis from thousands of genes to few hundreds of pathways. Second, identifying significant pathways is more meaningful than a list of different gene expression when comparing two samples such as normal and cancerous.

Pathway analysis has useful applications such as discovering disease occurrences by finding out the disrupted biological pathways. Another application is drugs development that aims to design a drug that target one or two disrupted pathways [1,28]. Moreover, researchers plan to use biological pathway approach to personalized patients treatment and drug development [1,28].

The paper is organized as follows. Section 2 overviews pathway databases. Gen expression, microarray and RNA-seq are presented in Sect. 3. Section 4 overviews pathway analysis techniques. Section 5 describes some miRNA analysis techniques. The conclusion is presented in Sect. 6.

2 Pathway Databases

Pathways are curated manually from biological experiments or automatically using text mining techniques [24]. Manual curation is more accurate and reliable.

There are 547 available biological pathways related resources [2]. For example, KEGG (Kyoto Encyclopedia of Genes and Genomes) database is the most popular pathways resource. It contains manually curated and inferred metabolic, signaling and disease pathways for over 650 organisms [7]. Reactome is another example for manually curated and inferred pathways database. It contains metabolic, signaling and disease pathways for human [7]. Also, BioCarta contains manually curated metabolic and signaling pathways for human and mouse [7].

3 Gene Expression

Genes control cells functions and all cells have the same genetic information. Genes are active or inactive (have different gene expression) according to a cell type and different conditions. Gene expression measures amount of mRNA produced in a cell [13] and gives the degree to which gene is active under different conditions.

3.1 Microarray vs RNA-seq

Sequencing techniques allow scientists to analyze tens of thousands of genes in parallel at any given time [12]. These technologies help us to understand diseases and provide better treatments [5]. Sequencing techniques start with using microarray technologies. Then, next generation sequencing was developed and it has a lot of sequencing that used in gene expression analysis such as RNA-seq.

Microarray is a small glass or plastic or silicon chip in which tens of thousands of DNA molecules (probes) are attached. Microarray is able to detect specific DNA molecules of interest. It works as follow: from two mRNA samples (a test sample and a control sample) cDNAs are obtained and labelled with fluorescent dyes and then hybridized on the surface of the chip. Then, the chips are scanned to read the signal intensity that is omitted from the labelled and hybridized targets [12,23].

RNA-seq is used to rapid profile mRNA expression of whole transcriptome [5,9]. It works as follow: small reads are aligned to an annotated reference mRNA. Then, the number of reads that aligned to one of different cDNAs are counted [4]. RNA-seq outperforms microarray in various aspects. First, the ability of detecting and identifying unknown genes and detecting differential expression levels that have not detected by microarray [5]. Second, it does not require specific probes or predefined transcriptome of interest [5]. Third, it increases specificity and sensitivity for detecting genes [5].

Gene expression usually presents by $i \times j$ matrix as in Fig. 1 where the rows represent expression pattern of genes and the columns represent different conditions such as different samples (normal vs cancer) or different time points [12,13]. In microarray, x_{ij} represents intensity level of hybridization of ith gene in a jth condition. While in RNA-seq, x_{ij} represents the number of reads of gene i observed in condition j.

$$\begin{pmatrix} x_{1,1} & x_{1,2} & \cdots & x_{1,j} \\ x_{2,1} & x_{2,2} & \cdots & x_{2,j} \\ \vdots & \vdots & \ddots & \vdots \\ x_{i,1} & x_{i,2} & \cdots & x_{i,j} \end{pmatrix}$$

Fig. 1. Gene expression matrix

In microarray, chip is scanned to get hybridization data that are usually represent in a spreadsheet-like format [5] where each cell represents the intensity of hybridization of a specific gene in a specific condition as in Fig. 2. In RNA-seq, sequencing is used to get read counts that represent in spreadsheet-like format. Each cell represents the number of reads that aligned to one of thousands of different cDNAs [5] as in Fig. 3.

Having gene expression available a lot of analysis techniques can be applied. Pathway analysis is one of them that have impact on the development of drugs and disease diagnosis.

4 Classification of Pathway Analysis Techniques

Pathway analysis can be classified into two approaches: detecting significant pathways and discovering new pathways as in Fig. 4. Detecting significant

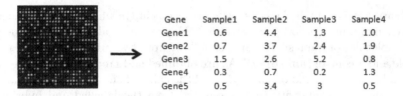

Fig. 2. Microarray gene expression matrix

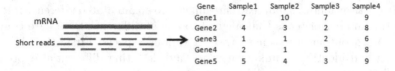

Fig. 3. RNA-seq gene expression matrix

pathways approach aims to define and rank significant pathways that related to a specific phenotype either by enrichment score analysis or machine learning techniques [18]. Shin and Kim [23] classified the computational approaches for pathways analysis into three groups: clustering-based methods, gene-based methods and gene set-based methods. Clustering based methods are based on assumption that genes with similar expression would have similar functions or involved in the same biological processes [13,23]. Therefore, genes are clustered and pathways for each cluster are determined. In gene-based methods, differentially expressed genes DEGs between two samples (a test sample and a control sample) are identified, and then significant pathways that DEGs are involved are determined. In gene set-based methods the gene expression and a prior biological resource (i.e. pathway databases) are used to determine the significant pathways (gene sets) [23].

Discovering new pathways can be achieved either by mining the literature through text mining techniques or automatic inferring pathways from network interactions or gene expression data. In this paper, we focus on detecting significant pathways approaches and we categorize the research in the area according to the type of gene expression to be analyzed into two categories: pathway-based microarray analysis, and pathway-based RNA-seq analysis. Next we will review some research related to each category.

4.1 Pathway-Based Microarray Analysis

Most research are focused on analyzing microarray gene expression either to determine significant pathways that contribute to a phenotype of interest or deal with features (genes) selection problem. Next some research related to classification, feature selection and clustering approaches are reviewed.

Classification. It aims to define and rank significant pathways that related to a specific phenotype using machine learning approaches. Zhang et al. [29] used

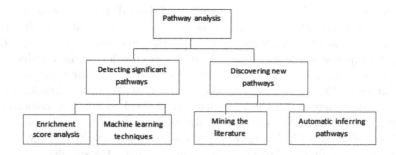

Fig. 4. classification of pathway analysis techniques

machine learning algorithms: nave bayes, support vector machine, decision tree and random forests to rank pathways based on classification error. By using three microarray expression datasets, they proved that machine learning algorithms outperform enrichment score analysis in identifying significant pathways. Pang et al. [20] used random forest classification and regression to analyze and rank pathways. In addition, they pointed out that their method was the first that used continuous measures for ranking pathways.

Features Selection. It aims to select informative genes within pathways before the pathway evaluation process to reduce computational time and improve accuracy [19]. Misman et al. [19] pointed out that when observing a particular biological context such as cancer some genes within pathways are only responsible for a phenotype. Thus, selecting subset of genes is important phase before ranking pathways. Zhang et al. [29] used minimum redundancy maximum relevance mRMR to select representative genes from each pathway. Panteris et al. [22] selected significant genes from each pathway (pathway signature) that describe the pathway at a given experimental condition. Misman et al. [19] used SVM-SCAD to select genes within pathways and have used B-type generalized approximate cross validation (BGACV) to select appropriate tuning parameter for SVM-SCAD. Jungjit et al. [14] proposed a KEGG pathway-based feature selection method for multi-label classification. Their method selects genes based on weighted formula that combines genes predictive accuracy and their occurrence in cancer-related KEGG pathways. Ibrahim et al. [11] selected strongly correlated genes for accurate disease classification by using pathways as prior knowledge. Their method was compared with five feature selection methods using two classifiers: K-nearest neighbour and support vector machine and it preformed the best for three microarray datasets.

Clustering. Detecting pathways in clustering analysis is used as a validation measure or as a partitioning measure. The reason of validation measure is to prove the validity of a clustering algorithm and for partitioning measure to partition datasets into biological meaningful clusters. For example, Shin and

Kim [23] used hierarchical clustering with Euclidean distance to generate gene clusters from gene expressions. Then, pathways are identified in each cluster to check the validity of clustering to identify the subclasses of leukemia. Zhao et al. [30] proposed a pathway-based clustering approach that used pathways to identify clusters. Their aim was to identify subgroups of cancer patients that may respond to the same treatment. Since cancers have similar phenotypes but resulting from different genetic mutations which lead to different responses to the same treatment. Their method is as follow: identify differential gene expression. Then, identify KEGG pathways that enriched with DGEs. Finally, classify the samples according to the expression of genes within the specified pathways. Also, Milone et al. [17] proposed a new method based on self-organizing map SOM clustering that used common metabolic pathways and Euclidean distance as similarity measures to construct clusters. Their objective was to improve the quality of clustering formation by combining pathway information. They used transcripts and metabolites datsets form Solanum lycopersicum and Arabidopsis thaliana species. Their method just improved the biological meaning of clusters compared with classical SOM. Moreover, Kozielski and Gruca [16] proposed a method that combined gene expression and gene ontology to identify clusters. So, the cluster membership should satisfy both gene expression and gene ontology. The proposed method is based on fuzzy clustering algorithm. Pang and Zhao [21] have proposed a method to generate pathways clusters that are related to a phenotype of interest from pathway- based classification [20]. They used class votes from random forest as similarity measure between pathways and tight clustering approach. Table 1 summarizes the research in pathway-based clustering and explains dataset, aim of clustering and aim of pathway analysis either validation or partitioning measure.

4.2 Pathway-Based RNA-seq Analysis

There are limited research focusing on analyzing RNA-seq gene expression to determine significant pathways that contribute to a phenotype of interest. Theses research focusing on statistical approaches. For example, Xiong et al. [27] developed a tool set that have multiple gene-level and gene set-level statistics to determine significant pathways. Fridley et al. [9] proposed using gamma method with soft truncation threshold to determine the gene sets that related to particular phenotype. Then, they applied the method to a smallpox vaccine immunogenetic study to identify gene sets or pathways with differential expression genes between high and low responders to the vaccine. Wang and Cairns [26] proposed combining differential expression with splicing information to detect significant gene sets based on Kolmogorov-Smirnov-like statistic. Xiong et al. [27] pointed out that the Wang and Cairns method is computationally expensive. Also, Hanzelmann et al. [10] developed a method that calculates variation of pathway activity profile over a sample population to analyze gene sets. Their method can be applied to RNA-seq as well as microarray data.

Table 1. Clustering and pathway analysis

Ref	Clustering algorithm	Dataset	Aim of clustering	Aim of pathway analysis
[23]	Hierarchical clustering	Leukemia	To identify subtypes of leukemia	Validation measure
[30]	Hierarchical clustering	NCI60 and DLBCL datasets	To identify subtypes of cancers	Partitioning measure
[17]	Self-organizing map SOM	Solanum lycopersicum and Arabidopsis thaliana	To improve the biological meaning of clusters	Partitioning measure
[21]	Tight clustering approach	Pathways from KEGG, BioCarta and Gen-Mapp	To generate pathways clusters that are related to a phenotype	———

5   Pathway-Based MiRNA Analysis

There are few research focusing on analyzing miRNA to determine significant pathways. Among them, Chen et al. [8] proposed a mathematical model (Bayesian implementation) that used miRNA targets for mapping miRNA to pathways then applied hypothesis test to extract significant pathways. Zhang et al. [28] used sample-matched miRNA and mRNA expression and pathway structure to analyze glioma patient survival. Wang et al. [25] suggested using functional information (gene ontology) to improve miRNA target prediction algorithms since genes that regulated by the same miRNA may share similar functions. Most miRNa target prediction algorithms used physical interaction mechanisms such as free energy, seed match and sequence conservation. Wang et al. [25] built SVM ensemble classifier that combined gene ontology and sequence information to predict miRNA targets.

6   Conclusion

Pathway analysis is reliable in discovering diseases and has various useful applications. In addition, data mining techniques are applied to pathway analysis to discover biological interesting hidden information. Most research in the field are based on analysis of microarray datsets and few are based on RNA-seq. Thus, applying data mining approaches such as classification and clustering to pathway-based RNA-seq analysis leads to more biological results.

References

1. Biological pathways fact sheet (2014). http://www.genome.gov/27530687. Accessed 11 August 2014
2. Pathguide (2015). http://www.pathguide.org/. Accessed 02 January 2015
3. Pathway analysis (2014). http://www.genexplain.com/pathway-analysis. Accessed 08 November 2014
4. Getting started with RNA-seq data analysis (2011). http://www.illumina.com/documents/products/datasheets/datasheet_rnaseq_analysis.pdf
5. Transitioning from microarrays to mRNA-seq, December 2011. http://www.illumina.com/content/dam/illumina-marketing/documents/icommunity/article_2011_12_ea_rna-seq.pdf
6. American cancer society: cancer facts and figures 2014 (2014)
7. Carugo, O., Eisenhaber, F.: Data Mining Techniques for the Life Sciences. Springer, New York (2010)
8. Chen, Y., Chen, H.I., Huang, Y.: Mapping miRNA regulation to functional gene sets. In: International Joint Conference on Bioinformatics, Systems Biology and Intelligent Computing, IJCBS 2009, pp. 122–125. IEEE (2009)
9. Fridley, B.L., Jenkins, G.D., Grill, D.E., Kennedy, R.B., Poland, G.A., Oberg, A.L.: Soft truncation thresholding for gene set analysis of RNA-seq data: application to a vaccine study. Sci. Rep. **3**, 2898 (2013)
10. Hänzelmann, S., Castelo, R., Guinney, J.: GSVA: gene set variation analysis for microarray and RNA-seq data. BMC Bioinf. **14**(1), 7 (2013)
11. Ibrahim, M.H., Jassim, S., Cawthorne, M., Langlands, K.: Pathway-based gene selection for disease classification. In: 2011 International Conference on Information Society (i-Society), pp. 360–365. IEEE (2011)
12. Jiang, D., Tang, C., Zhang, A.: Cluster analysis for gene expression data: a survey. IEEE Trans. Knowl. Data Eng. **16**(11), 1370–1386 (2004)
13. Jones, N.C., Pevzner, P.: An Introduction to Bioinformatics Algorithms. MIT press, Cambridge (2004)
14. Jungjit, S., Michaelis, M., Freitas, A.A., Cinatl, J.: Extending multi-label feature selection with KEGG pathway information for microarray data analysis. In: 2014 IEEE Conference on Computational Intelligence in Bioinformatics and Computational Biology, pp. 1–8. IEEE (2014)
15. Khatri, P., Sirota, M., Butte, A.J.: Ten years of pathway analysis: current approaches and outstanding challenges. PLoS Comput. Biol. **8**(2), e1002375 (2012)
16. Kozielski, M., Gruca, A.: Soft approach to identification of cohesive clusters in two gene representations. Procedia Comput. Sci. **35**, 281–289 (2014)
17. Milone, D.H., Stegmayer, G., López, M., Kamenetzky, L., Carrari, F.: Improving clustering with metabolic pathway data. BMC Bioinf. **15**(1), 101 (2014)
18. Misman, M., Deris, S., Hashim, S., Jumali, R., Mohamad, M.: Pathway-based microarray analysis for defining statistical significant phenotype-related pathways: a review of common approaches. In: International Conference on Information Management and Engineering, ICIME 2009, April 2009, pp. 496–500 (2009)
19. Misman, M.F., Mohamad, M.S., Deris, S., Abdullah, A., Hashim, S.Z.M.: An improved hybrid of SVM and SCAD for pathway analysis. Bioinformation **7**(4), 169 (2011)
20. Pang, H., Lin, A., Holford, M., Enerson, B.E., Lu, B., Lawton, M.P., Floyd, E., Zhao, H.: Pathway analysis using random forests classification and regression. Bioinformatics **22**(16), 2028–2036 (2006)

21. Pang, H., Zhao, H.: Building pathway clusters from random forests classification using class votes. BMC Bioinf. **9**(1), 87 (2008)
22. Panteris, E., Swift, S., Payne, A., Liu, X.: Mining pathway signatures from microarray data and relevant biological knowledge. J. Biomed. Inf. **40**(6), 698–706 (2007)
23. Shin, M., Kim, J.: Data mining and knowledge discovery in real life applications. In: Microarray Data Mining for Biological Pathway Analysis, pp. 319–336. I-Tech (2009)
24. Viswanathan, G.A., Seto, J., Patil, S., Nudelman, G., Sealfon, S.C.: Getting started in biological pathway construction and analysis. PLoS Comput. Biol. **4**(2), e16 (2008)
25. Wang, N., Wang, Y., Yang, Y., Shen, Y., Li, A.: miRNA target prediction based on gene ontology. In: 2013 Sixth International Symposium on Computational Intelligence and Design (ISCID), vol. 1, pp. 430–433. IEEE (2013)
26. Wang, X., Cairns, M.J.: Gene set enrichment analysis of RNA-seq data: integrating differential expression and splicing. BMC Bioinf. **14**(Suppl. 5), S16 (2013)
27. Xiong, Q., Mukherjee, S., Furey, T.S.: GSAASeqSP: a toolset for gene set association analysis of RNA-seq data. Sci. Rep. **4**, 6347 (2014)
28. Zhang, C., Li, C., Li, J., Han, J., Shang, D., Zhang, Y., Zhang, W., Yao, Q., Han, L., Xu, Y., Yan, W., Bao, Z., You, G., Jiang, T., Kang, C., Li, X.: Identification of miRNA-mediated core gene module for glioma patient prediction by integrating high-throughput miRNA, mRNA expression and pathway structure. PLoS ONE **9**(5), e96908 (2014)
29. Zhang, W., Emrich, S., Zeng, E.: A two-stage machine learning approach for pathway analysis. In: 2010 IEEE International Conference on Bioinformatics and Biomedicine (BIBM), December 2010, pp. 274–279 (2010)
30. Zhao, X., Zhong, S., Zuo, X., Lin, M., Qin, J., Luan, Y., Zhang, N., Liang, Y., Rao, S.: Pathway-based analysis of the hidden genetic heterogeneities in cancers. Genomics, Proteomics Bioinf. **12**(1), 31–38 (2014)

Aspects of Data Mining

Identify Error-Sensitive Patterns
by Decision Tree

William Wu[✉]

QCIS FEIT, University of Technology Sydney, Ultimo, NSW, Australia
will.edu.au@gmail.com

Abstract. When errors are inevitable during data classification, finding a particular part of the classification model which may be more susceptible to error than others, when compared to finding an Achilles' heel of the model in a casual way, may help uncover specific error-sensitive value patterns and lead to additional error reduction measures. As an initial phase of the investigation, this study narrows the scope of problem by focusing on decision trees as a pilot model, develops a simple and effective tagging method to digitize individual nodes of a binary decision tree for node-level analysis, to link and track classification statistics for each node in a transparent way, to facilitate the identification and examination of the potentially "weakest" nodes and error-sensitive value patterns in decision trees, to assist cause analysis and enhancement development.

This digitization method is not an attempt to re-develop or transform the existing decision tree model, but rather, a pragmatic node ID formulation that crafts numeric values to reflect the tree structure and decision making paths, to expand post-classification analysis to detailed node-level. Initial experiments have shown successful results in locating potentially high-risk attribute and value patterns; this is an encouraging sign to believe this study worth further exploration.

1 Introduction

The ultimate goal of this study is to find the most problematic and error-sensitive part of a classification model, and this would require the collection, identification and comparison of classification statistics of its individual component parts. Decision trees have been selected as the pilot model for this study because it is a well-researched classification model with a simple structure, decisions on attributes and values are clearly displayed in a form of branches and nodes, as shown in Fig. 1.

Using the first branch of the above decision tree as an example:

alm1 <= 0.57 and aac <= 0.64 and mcg <= 0.73: neg (189.0/6.0)

This branch contains three nodes with three split point values, (1) <= 0.57 for attribute "alm1", (2) <= 0.64 for attribute "aac", and (3) <= 0.73 for attribute "mcg". While all these three split points play a role in leading to the 6 classification errors amongst the 189 instances along this classification path in the form of a decision tree branch, one key question is, which node and its related attribute and value may have

© Springer International Publishing Switzerland 2015
P. Perner (Ed.): ICDM 2015, LNAI 9165, pp. 81–93, 2015.
DOI: 10.1007/978-3-319-20910-4_7

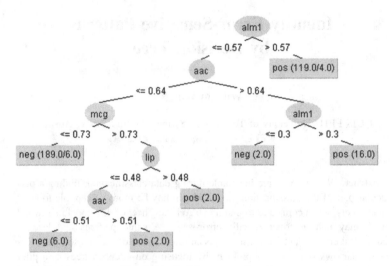

Fig. 1. A decision tree example

been more susceptible to the 6 errors? When expanding this node-specific examination to all branches of the decision tree model, the question can then generalized as – are there some tree nodes more error-prone and error-sensitive than others? If so, can the most error-sensitive nodes and their related attributes and values be identified in a systematic way? These two questions are now the focus of this study.

The rest of this paper is organized as follows. Section 2 describes some initial questions and thoughts that led to the decision tree digitization idea. Section 3 reviews some early influential work that inspired this study. Section 4 outlines the major steps in the decision tree digitization process. Section 5 summarizes experiments on five datasets. Section 6 discusses the experiment results and their implications. Finally, Sect. 7 concludes the current progress and outlines a plan for future exploration.

2   Initial Issues and Background

Decision trees provide an easy-to-follow graphical view of the classification process at a glance, outlines each classification rule from root to leaf step by step in the form of node by node. One issue with such visual representation is, when a large dataset is used and the decision tree structure becomes complex, its graphical view can be clustered and muddled by the full-blown mass of crisscross branches and nodes even when pruning is applied. It can obscure the identification of attributes and values when detailed analysis is required on certain classification rules and components.

Another issue is, when node-level statistics are required in such a detailed analysis, the visual space reserved for each node on a decision tree may not be the most suitable place to present its node-level statistic values, as this would cluster the presentation of existing branches and nodes even further, making the node scanning and visual interpretation process even more difficult.

One possible solution to address these issues is, to provide a unique tag for each tree node, to collect and maintain the node-level classification statistics away from the tree structure, and to link them with their respective node by using the unique node tag as the retrieval key. As a result, classification statistics of each node can be stored and analyzed without any convoluted addition to the existing tree structure.

This decision tree digitization method and error-sensitive pattern analysis may seem trivial when compared to some major published work as outlined in Sect. 3, but because no such specific analysis has been observed so far, that nevertheless inspired this study and the digitization idea development.

3 Related Work

Decision tree classification and attribute selection methodologies are two key research areas closely related to this study. Amongst the vast number of research literature and many innovative algorithms on decision trees, the C4.5 model [1] and CART model [2] are two benchmarks used as the foundation and guidance for the proposed decision tree digitization method. While the C4.5 model utilizes the gain ratio to "divide and conquer" data attributes and values to form a classification tree, the CART model makes use of the Gini impurity measure to split the attributes and data values to build a decision tree.

In the area of attribute selection, many of the renowned methodologies, such as Information Gain, Gain Ratio, Gini Index, RELIEF, SFS and SBE algorithm [1–6], as well as some newly established techniques such as SAGA [7] and UFSACO [8], they have been developed as a pre-process procedure or as an integrated part of the classification process. One key logic shared by these algorithms is to select and prioritize the most informative and differentiating data attributes before or during classification. While these methodologies have improved the classification performance in a holistic and "macro" way, they are not particularly designed to examine attribute and value patterns at an individual node and "micro" level.

Ongoing research and development have resulted new techniques in data sampling and classification process, such as bagging [9], boosting [10] and randomization to reduce bias [11], and provided the ground for individual tree models to be incorporated into an ensemble of tens or hundreds or even thousands of classifiers to achieve better performance. For example, the AdaBoost model [12] that adapts a weak learning algorithm such as a decision tree model as a starting point, then to reweight samples, retrain and rebuild a new tree after each intermediate learning cycle, and to vote in the best performing tree from "the crowd". RandomForest [13] is another popular ensemble approach; it applies the bagging technique to a subset of attributes as well as training samples that are randomly selected, to generate new trees via iterations and to vote in the best performing tree amongst the peers in "the forest". The AdaTree method [14], the Probabilistic Boosting-Tree method [15], and a combined Bayesian model approach [16] are some new additions to the ensemble trend development.

Compared to standalone decision tree models, ensemble methods provide higher accuracy but at the cost of increased complexity and computational resources, the clarity and ease of result interpretation have also been reduced [17, 18]. Amongst the above and other major literatures that have been studied, detailed analysis on error-sensitive attributes and values had not been apparent. One recent evaluation study on error-sensitive attributes (ESA) [19] intended to begin a detailed examination at an attribute level based on three specific terms using binary decision trees. The term "ambiguous value range" describes the "overlapped" value ranges between Positive and Negative instances; the term "attribute-error counter" describes the number of misclassified instances of an attribute with attribute values reside in ambiguous value ranges; and the term "error-sensitive attribute" describes an attribute that is considered to be more prone and risky to cause or associate with errors in a data classification process. All the above work has provided either the inspiration or the basis for the current study, and explained why decision trees have been adopted as the pilot model in this initial phase.

4   The Decision Tree Digitization Process

A decision tree model can be transformed into an array of branches and each branch consists of an array of nodes, and each node represents its underlying attribute and value's split point condition. Because each node can be considered as a child node from its immediate parent node, and all levels of parent nodes can be traced back to the root node as the origin, therefore, each node can be uniquely identified by a form of regression or inference process based on its hierarchical position within the tree and using the root as the starting point, and a graphical tree can subsequently be mapped into a matrix of referential and digitized node IDs, which can link and retrieve node level classification statistics for detailed analysis. The following pseudo code outlines the digitization process step-by-step:

Input: a binary decision tree with m branches and each branch contains a varying number of nodes, and a dataset with n instances

Output: an enumerated map of individual node IDs and a collection of node level classification statistics

Process: stage-1 is to enumerate each tree node and produce a map of node IDs; stage-2 is to collect classification statistics for individual nodes and using IDs as keys

Stage-1: construct an enumerated decision tree map

>   for 1 to m branches of the decision tree
>   for all nodes in the current branch

1. if a node is the root node then assign "1" as the starting value of its node ID
2. if a node is an immediate child node from the root node, then first append a "." to the current node ID as a node delimiter, then add x to form 1.x as its 2nd part of the node ID, x denotes the current number of immediate child nodes branching out from the root and increments by 1 by counting from left to right, e.g. 1.1 as the 1st child node, 1.2 as the 2nd child node, and so on
3. if a non-root node has child nodes then first append a "." to the current node ID as a node delimiter, then assign 1 to its 1st child node on the left as its node ID, assign 2 to its 2nd child node on the right as its node ID; a node ID example is: 1.1.2.2.1

Stage-2: traverse and collect individual node level classification statistics

for 1 to *n* data instances
for 1 to *m* branches in the enumerated matrix map
for all nodes in the current branch

1. if current instance's attribute value satisfies current node's split point value condition, then continues to next node along current branch
2. if current node's split point condition cannot be satisfied, then advance to the start of next branch in the map
3. if end of current branch is reached and the leaf-node condition is satisfied, then update and store the node-level statistics using the node ID as the key for all nodes of the current branch, and move to the next data instance

On completion of the tree digitization and statistics collection, a simple ranking of the classification error rate by node IDs can then potentially reveal the "weakest" and most error-sensitive node in the tree. The word "potentially" has to be highlighted and emphasized here. Using a node's error rate value instead of its error count number may avoid the bias towards "heavy traffic" nodes; however, this may unduly magnify the "weakness" of some "low traffic" nodes. For example, node-A has been traversed by 10 instances with 5 errors and its error rate is 50 %, node-B has been traversed by 100 instances with 48 errors and its error rate is 48 %, while node-A is subsequently ranked as a "weaker" node than node-B by comparing error rate, this may not necessarily be true when more data are used for testing. In a later stage of the study, significant test and threshold value control on selection criteria can be implemented as an enhancement measures.

Nevertheless, this decision tree digitization method is another step forward in the study of error-sensitive value patterns in data classification, and results of initial experiments appeared to be supporting this idea.

5 Experiments

During the evaluation study of error-sensitive attributes (ESA) [19], five UCI datasets [20], Ecoli, PIMA Diabetes, Wisconsin Cancer, Liver Disorder and Page Blocks, were used in the evaluation process. These datasets have been used again in the current study so their experiment results can be analyzed and compared side by side against the ESA evaluation results. Experiments have been conducted using WEKA's [21] C4.5/J48 decision tree classifier with standard configuration, e.g. confidence factor for pruning is 0.25, minimum number of instances per leaf is 2, MDL correction is used and test option is 10-fold cross-validation.

5.1 Digitization Reflects Decision Trees in a Concise and Effective Way

A decision tree model for the Ecoli dataset contains 7 branches and 13 nodes, as shown in Fig. 1. On completion of its digitization, the digitized form of branches and nodes is shown in the 1st row of Table 1. Each node is uniquely tagged by a digital ID, and each ID reflects the node's hierarchical location in the tree. Because of its self-structured and self-referenced nature, the ID also encapsulates its preceding nodes of the same branch and presents itself as a compact and enumerated decision path; therefore, a collective display of each branch's leaf node resembles the decision tree model in a simplified and digitized form, as shown in the 2nd row of Table 1.

Table 1. Ecoli dataset's decision tree in digitized form

Numerated tree map showing all node IDs in each decision path	Branch 1: 1.0 -> 1.1 -> 1.1.1 -> **1.1.1.1** Branch 2: 1.0 -> 1.1 -> 1.1.1 -> 1.1.1.2 -> 1.1.1.2.1 -> **1.1.1.2.1.1** Branch 3: 1.0 -> 1.1 -> 1.1.1 -> 1.1.1.2 -> 1.1.1.2.1 -> **1.1.1.2.1.2** Branch 4: 1.0 -> 1.1 -> 1.1.1 -> 1.1.1.2 -> **1.1.1.2.2** Branch 5: 1.0 -> 1.1 -> 1.1.2 -> **1.1.2.1** Branch 6: 1.0 -> 1.1 -> 1.1.2 -> **1.1.2.2** Branch 7: 1.0 -> **1.2**
Leaf-node IDs resemble a simplified and enumerated tree	Branch 1: **1.1.1.1** Branch 2: **1.1.1.2.1.1** Branch 3: **1.1.1.2.1.2** Branch 4: **1.1.1.2.2** Branch 5: **1.1.2.1** Branch 6: **1.1.2.2** Branch 7: **1.2**

In the second example, the Pima diabetes dataset model has 20 branches and 39 nodes, as shown in the left column of Table 2, they have been concisely and effectively represented by their leaf-node IDs, as shown in the right column of Table 2:

Table 2. Pima dataset's decision tree represented by enumerated leaf-node IDs

Pima diabetes dataset's decision tree model	Leaf-node IDs
	Branch 1: 1.1.1 Branch 2: 1.1.2.1 Branch 3: 1.1.2.2.1 Branch 4: 1.1.2.2.2.1 Branch 5: 1.1.2.2.2.2.1.1 Branch 6: 1.1.2.2.2.2.1.2.1 Branch 7: 1.1.2.2.2.2.1.2.2.1 Branch 8: 1.1.2.2.2.2.1.2.2.2 Branch 9: 1.1.2.2.2.2.2 Branch 10: 1.2.1.1 Branch 11: 1.2.1.2.1 Branch 12: 1.2.1.2.2.1.1 ... Branch 18: 1.2.2.1.2.1 Branch 19: 1.2.2.1.2.2 Branch 20: 1.2.2.2

Similarly, the Wisconsin cancer dataset's decision tree which has 14 branches and 27 nodes, the Liver Disorders dataset's decision tree which has 26 branches and 51 nodes, and the Page Blocks dataset's decision tree which has 41 branches and 81 nodes, all have been correctly mapped by their leaf-node IDs respectively in a concise manner, as shown in Table 3:

Table 3. Another 3 decision trees represented by enumerated lead-node IDs

Wisconsin cancer dataset's enumerated decision tree	Liver Disorders dataset's enumerated decision tree	Page Blocks dataset's enumerated decision tree
Branch 1: 1.1.1	Branch 1: 1.1.1.1	Branch 1: 1.1.1.1
Branch 2: 1.1.2.1	Branch 2: 1.1.1.2.1	Branch 2: 1.1.1.2
Branch 3: 1.1.2.2.1.1	Branch 3: 1.1.1.2.2.1.1	Branch 3: 1.1.2.1.1.1
Branch 4: 1.1.2.2.1.2	Branch 4: 1.1.1.2.2.1.2.1	...
Branch 5: 1.1.2.2.2	Branch 5: 1.1.1.2.2.1.2.2.1	Branch 12: 1.1.2.2.2
Branch 6: 1.2.1.1	Branch 6: 1.1.1.2.2.1.2.2.2	Branch 13: 1.2.1
Branch 7: 1.2.1.2	...	Branch 14: 1.2.2.1.1.1.1.1.1
Branch 8: 1.2.2.1.1.1	Branch 17: 1.2.1.1.2.1.2.2.1	...
Branch 9: 1.2.2.1.1.2	Branch 18: 1.2.1.1.2.1.2.2.2	Branch 30: 1.2.2.1.2.1.1.2.1.2.1
Branch 10: 1.2.2.1.2.1.1	Branch 19: 1.2.1.1.2.2.1	Branch 31: 1.2.2.1.2.1.1.2.1.2.2
Branch 11: 1.2.2.1.2.1.2.1	Branch 20: 1.2.1.1.2.2.2.1	Branch 32: 1.2.2.1.2.1.1.2.2
Branch 12: 1.2.2.1.2.1.2.2	...	...
Branch 13: 1.2.2.1.2.2	Branch 24: 1.2.2.1.2.1	Branch 39: 1.2.2.2.2.1
Branch 14: 1.2.2.2	Branch 25: 1.2.2.1.2.2	Branch 40: 1.2.2.2.2.2.1
	Branch 26: 1.2.2.2	Branch 41: 1.2.2.2.2.2.2.2

Once each individual node is uniquely tagged, its classification statistics can then be collected, stored and analyzed at an individual node- and micro-level, as compared to the typical holistic model- and macro-level analysis based on the whole decision tree and its overall result. Some of the statistics collection and comparison results are documented in the following section.

5.2   Node-Level Statistics Comparison and Error-Sensitive Pattern Identification

There may be different ways to examine the node-level statistics, for example, to compare the "heaviest" and "lightest" nodes using the highest and lowest counts of instances that traversed through, but the focus of this study is to identify and explore the "weakest" nodes with the highest error rates and the involved value patterns.

As a first step, the attributes and values involved with the top-3 nodes in error rate ranking are compared with the top-3 ranked attributes identified in the error-sensitive attribute (ESA) evaluation [19], to cross-check these two different error-sensitive pattern evaluation methods, as shown in Table 4. It is showing that three datasets - Pima, Wisconsin and Page Blocks, have closely comparable "underscored" error-sensitive attributes, and two datasets - Ecoli and Liver Disorders, have partially comparable "underscored" error-sensitive attributes.

Table 4. The "weakest" nodes' attributes & values VS. The most error-sensitive attributes

Rank	Ecoli dataset's enumerated tree node & error-rate ... attributes & values involved	Attributes identified in ESA evaluation by attribute-error count
1	**1.2** (3.36%: 4/119) ... <u>alm1</u>>0.57	chg (17)
2	**1.1.1.1** (3.17%: 6/189) ... <u>alm1</u><=0.57 & aac<=0.64 & mcg<=0.73	<u>alm1</u> (15)
3	**1.1.1** (3.02%: 6/199) ... <u>alm1</u><=0.57 & aac<=0.64	lip (14)

Rank	Pima diabetes dataset's enumerated tree node & error-rate ... attributes & values involved	Attributes identified in ESA evaluation by attribute-error count
1	**1.1.2.2.2.1** (40.48%: 34/84) ... <u>plas</u><=127 & <u>mass</u>>26.4& <u>age</u>>28 & plas>99 & pedi<=0.561	<u>plas</u> (83)
2	**1.2.2.1.2.1** (32.50%: 13/40) ... <u>plas</u>>127 & <u>mass</u>>29.9 & <u>plas</u><=157 & pres>61 & age<=30	nmass (70)
3	**1.1.2.2.2** (30.51%: 36/118) ... <u>plas</u><=127 & <u>mass</u>>26.4 & <u>age</u>>28 & plas>99	age (31)

Rank	Wisconsin cancer dataset's enumerated tree node & error-rate ... attributes & values involved	Attributes identified in ESA evaluation by attribute-error count
1	**1.2.2.1.2.1.2.1** (20.00%: 1/5) ... UC_Sz>2 & UC_Sh>2 & UC_Sz<=4 & Bare_Nuc>2 & Clump_Th<=6 & UC_Sz>3 & Mg_Adh<=5	<u>UC_Sz</u> - Unif Cell Size (35)
2	**1.2.2.1.2.1.1.1** (15.38%: 2/13) ... UC_Sz>2 & UC_Sh>2 & UC_Sz<=4 & Bare_Nuc>2 & Clump_Th<=6 & UC_Sz<=3	<u>UC_Sh</u> - Unif Cell Shape (35)
3	**1.2.2.1.2.1** (13.04%: 3/23) ... UC_Sz>2 & UC_Sh>2 & UC_Sz<=4 & Bare_Nuc>2 & Clump_Th<=6	<u>Clump_Th</u> - Clump Thickness (30)

Rank	Liver Disorders dataset's enumerated tree node & error-rate ... attributes & values involved	Attributes identified in ESA evaluation by attribute-error count
1	**1.2.2.1.2.2** (40.91%: 18/44)... gammagt>20 & drinks>5 & drinks<=12 & sgpt>21 & <u>sgot</u>>22	sgot (22)
2	**1.1.2.2.1.1** (38.10%: 8/21)... gammagt<=20 & sgpt>19 & <u>sgot</u>>20 & drinks<=5 & sgpt<= 26	mcv (16)
3	**1.2.2.1.2** (34.55%: 19/55)... gammagt>20 & drinks>5 & drinks<=12 & sgpt >21	alkphos (10)

Rank	Page Blocks dataset's enumerated tree node & error-rate ... attributes & values involved	Attributes identified in ESA evaluation by attribute-error count
1	**1.2.2.1.1.1.2.1.1.2** (30.00%: 3/10)... height>3 & <u>eccen</u>>0.25 & height <=27 & wb_trans<=7 & p_black<=0.178 & wb_trans>4 & blackpix <=20 & area<=108 & blackpix>7	<u>mean_tr</u> (89)
2	**1.1.2.1.1.1** (28.57%: 2/7)... height<=3 & <u>mean_tr</u>>1.35 & lenght<=7 & height<=2 & blackpix<=7	p_black (43)
3	**1.1.2.1.1.1.1.2** (25.00%: 1/4) ... height<=3 & <u>mean_tr</u>>1.35 & lenght> 7 & mean_tr<=4.08 & height<=1 & wb_trans<=2 & <u>mean_tr</u>>3.75	
3	**1.2.2.1.1.1.1.2.1.1** (25.00%: 3/12)... height>3 & <u>eccen</u>>0.25 & height<= 27 & wb_trans<=7 & p_black<=0.178 & wb_trans>4 & blackpix<= 20 & area <=108	eccen (29)

In a second step, data records associated with the "weakest" nodes identified by the digitization method are removed and a re-test is carried out, and another re-test is carried out on the datasets after certain most error-sensitive attributes are removed as specified in the ESA evaluation study [19]. Initial results confirm improved accuracy in all five datasets after the "weakest" records are removed, and one improved significantly, as shown in Table 5. Also outlined in this table are the ESA removal scenario retest results, three datasets return improved accuracy, and the other two return poorer accuracy, and further analysis on the results is discussed in the Sect. 6.

Table 5. Three-way performance comparison after removing the potentially "weakest" records and the most error-sensitive attributes

Ecoli's original dataset of 336 records	Re-test 217 records after removing 119 "weakest" records	Re-test original data after removing top most ESA – alm1
Accuracy: 94.05% with 20 errors	Accuracy: 97.70% with 5 errors ☑	Accuracy: 92.86% with 24 errors ☒
Pima diabetes' original dataset of 768 records	**Re-test 684 records after removing 84 "weakest" records**	**Re-test original data after removing top 2 most ESAs – plas & mass**
Accuracy: 73.83% with 201 errors	Accuracy: 77.19% with 156 errors ☑	Accuracy: 67.84% with 247 errors ☒
Wisconsin cancer's original dataset of 699 records	**Re-test 694 records after removing 5 "weakest" records**	**Re-test original data after removing top 2 most ESAs - UC_Sz & UC_Sh**
Accuracy: 94.13% with 41 errors	Accuracy: 95.97% with 28 error ☑	Accuracy: 95.71% with 30 errors ☑
Liver Disorders' original dataset of 345 records	**Re-test 301 records after removing 44 "weakest" records**	**Re-test original data after removing top 2 most ESAs – sgot & mcv**
Accuracy: 68.70% with 108 errors	Accuracy: 77.08% with 69 errors ☑	Accuracy: 71.01% with 100 errors ☑
Page Blocks' original dataset of 5473 records	**Re-test 5463 records after removing 10 "weakest" records**	**Re-test original data after removing top most ESA – mean_tr**
Accuracy: 97.19% with 154 errors	Accuracy: 97.36% with 144 errors ☑	Accuracy: 97.24% with 151 errors ☑

6   Experiment Analysis

The purpose of decision tree digitization is not simply to convert a graphical decision tree into a digital map of nodes, but rather, to use such a digital map to facilitate the collection of node-level statistics for the purpose of node-level error-sensitive value pattern analysis, to help highlight the potentially "weakest" part of the decision tree and the specific error-sensitive attributes and values involved, to distinguish data records with such risky value patters for further error analysis and the development of error-reduction measure. Results from the initial experiments appeared to be supporting this digitization idea and the identification of the "weakest" node and the related attribute and value patterns. The following sections discuss the results and possible implications.

6.1   Digitized Node IDs Facilitate Node-Level Analysis

The proposed decision tree digitization method makes the node-level analysis easy by formulating individual node IDs in a unique, numeric and contextual way. For example, the ID 1.2.1.1.2.1.2.2.1 as shown in the Liver Disorders example, is in a numeric text string format and incorporated with its preceding node IDs hierarchically within the same branch starting from the root. Because each ID is unique to the node in the tree, classification statistics can then be collected and stored for individual nodes using their IDs as the keys, and later to locate and retrieve the node-level statistics more efficiently than using the branch and node description text, e.g. *"gammagt > 20 & drinks <= 5 & drinks <= 3 & alkphos > 65 & sgot <= 24 & gammagt > 29 & mcv > 87*

& *mcv* <= *92*", even if such lengthy verbiage is consolidated and simplified as "*gammagt > 29 & drinks <= 3 & lkphos > 65 & sgot <= 24 & mcv between 87 and 92*", it is still awkward. As decision trees grow bigger, such concise node IDs can become more useful because of its systematic and self-referential characteristics.

One way to utilize such node-level statistics is, to ranking the classification error rate from high to low, the top node with the highest error rate may then be considered as the "weakest" node in the tree, and the attributes and values associated with the "weakest" node may be considered more error-sensitive than others, in relative terms.

6.2   Examine the "Weakest" Nodes and Error-Sensitive Value Patterns

To evaluate the effectiveness of the proposed digitization method, the subsequent node-level investigative results can be validated by performance comparison, and one practical but rather non-deterministic measure can be to compare the classification accuracy after some simplistic error-reduction measure is applied. For example, by using the attribute and value patterns associated with the "weakest" node, the potentially "weakest" and most error-sensitive data records, include both the misclassified and correctly classified data instances, can be identified and separated for further examination, and the original dataset becomes smaller in size but potentially higher in reliability and accuracy. Experiment results seemed to confirm the validity of the "weakest" node and the associated error-sensitive value patterns in all five datasets, one with a significant improvement in accuracy, and others with a modest but consistent level of improvement.

The best example is the Wisconsin cancer dataset with 699 records. After sorting and ranking the classification error rate of individual nodes, node 1.2.2.1.2.1.2.1 (20.00 %: 1/5) is identified as the "weakest" node, as shown in Table 4. When the five data records associated with this "weakest" node are identified and separated from the dataset, that is 5/699 = 0.007 % reduction in sample size, the accuracy improves by almost 2 % in a re-test, as shown in Table 5. Instead of one less error due to the removal of five error-sensitive records, there are 13 less errors in the re-test.

The other four datasets also show various levels of success in accuracy enhancement. For example, in the Liver Disorders dataset with 345 records, node 1.2.2.1.2.2 (40.91 %: 18/44) is identified as the "weakest" node, as shown in Table 4, and there are 44 records associated with this node and 18 of them are errors. A re-test to the updated dataset after the removal of those 44 error-sensitive records shows the actual error reduction is 39 instead of 18, and the overall accuracy has improved from 68.70 % in the original dataset to 77.08 % in the updated dataset.

One possible explanation to such impressive result is, the inclusions of the "weakest" data records have made "potentially significant" adverse impact to the info-gain (entropy) calculation when constructing C4.5/J48 decision trees because of their error-sensitive attributes and values, which leads to error-prone split point conditions and the consequent "weakest" nodes. If the impact is less significant, then the difference between the original and re-test result may be not so noticeable, as shown in the Page Blocks dataset.

This reasoning may partially explain why ensemble tree models, such as Random Forest, are considered superior to standalone tree models. The Random Forest model selects a portion of the data attributes randomly and generates hundreds and thousands of trees accordingly, and then votes for the best performing one to produce the classification result. The random attribute selection process may have inadvertently generated and voted for trees without some highly error-sensitive attributes, and also with bigger value ranges to split on due to fewer attributes involved, therefore enables ensemble models to produce more accurate results, but on the expense of resources and simplicity.

6.3   Discuss Possible Contribution, Effectiveness and Weakness

The evaluation study on error-sensitive attributes [19] has provided some constructive leads for this current study, but this decision tree digitization and node-level examination idea can be considered as another step forward because of its expansion from attribute level evaluation to individual node and split-value level evaluation. While still at an early stage, this latter expansion and study has shown encouraging and consistent experiment results, therefore, this can be considered as a potential contribution to the node-level analysis topic for decision trees.

In terms of effectiveness, this digitization method applies a digital way to tag each individual node of a decision tree uniquely and concisely with contextual reference, to simplify node-level statistics collection and analysis and expand the typical tree-level "macro" analysis with focus on the whole classification model into the node-level "micro" analysis with focus on specific attributes and values, and in a systematic and transparent way.

Meanwhile, the list of weakness of this study is also long and obvious. First, the successful experiments are based on the removal of the "weakest" data records, which may seem drastic and lacking of formal and theoretical proof; however, this has still highlighted the usefulness in identifying the "weakest" value patterns. This has led to the second major weakness - it is unclear what to do with the "weakest" records. Their removal improves overall accuracy, so a new question is, should a separate model be used to evaluate these error-sensitive records? If the "one size fits all" approach is not recommended, why not introduce a separate model for the "doubtful" data? The third major weakness is, this study is not based on ensemble methods, and ensemble trees are now the preferable classification models due to their superior performance to the standalone decision tree models, this makes the proposed decision tree digitization method less relevant to the latest classification development. Despite more weaknesses are still to be discussed, they have been recognized and will be used as a form of inspiration to broaden and advance this study.

7   Conclusion

This study attempts to address the question - "Is there a way to identify an Achilles' heel of a classification model?", that is, finding a way to locate the 'weakest' and most error-sensitive spot in the model. Towards this goal, the study develops a decision tree digitization method to facilitate the identification and examination of the potentially "weakest" nodes and error-sensitive value patterns in the model using decision trees as a pilot model. Initial experiment have demonstrated successful results when comparing to earlier evaluation study of error-sensitive attributes, but also prompted more questions.

Many of the study's own weak and questionable areas have been recognized, such as the need of formal and theoretical proof, the expansion of evaluation into ensemble methods and non-binary trees etc., and they will form the basis for the next phase of the study, such as a revision of the digitization method to cover ensemble models, and to find a more logical way to understand and utilize the "weakest" data records with error-sensitive value patterns.

References

1. Quinlan, J.R.: C4. 5: Programs for Machine Learning. Morgan Kaufmann, San Francisco (1993)
2. Breiman, L., Friedman, J.H., Olshen, R.A., Stone, C.J.: Classification and Regression Trees. Wadsworth International Group, Belmont (1984)
3. Alpaydin, E.: Introduction to Machine Learning. The MIT Press, London (2004)
4. Han, J., Kamber, M.: Data Mining: Concepts and Techniques. Morgan Kaufmann, San Francisco (2006)
5. Saeys, Y., Inza, I., Larranaga, P.: A review of feature selection techniques in bio-informatics. Bioinformatics 23(19), 2507–2517 (2007)
6. Witten, I.H., Frank, E.: Data Mining: Practical Machine Learning Tools and Techniques. Morgan Kaufmann, San Francisco (2005)
7. Gheyas, I.A., Smith, L.S.: Feature subset selection in large dimensionality domains. Pattern Recogn. 43(1), 5–13 (2010)
8. Tabakhi, S., Moradi, P., Akhlaghian, F.: An unsupervised feature selection algorithm based on ant colony optimization. Eng. Appl. Artif. Intell. 32, 112–123 (2014)
9. Breiman, L.: Bagging Predictors. Mach. Learn. 24(2), 123–140 (1996)
10. Schapire, R.E.: The Strength of Weak Learnability. Mach. Learn. 5(2), 197–227 (1990)
11. Ho, T.K.: Random decision forests. In: Proceedings of the Third International Conference on Document Analysis and Recognition, vol. 1, pp. 278–282 (1995)
12. Freund, Y., Schapire, R.E.: A decision-theoretic generalization of online learning and an application to boosting. In: Computational Learning Theory, pp. 23–37 (1995)
13. Breiman, L.: Random Forests. Mach. Learn. 45(1), 5–32 (2001)
14. Grossmann, E.: AdaTree: boosting a weak classifier into a decision tree. In: Computer Vision and Pattern Recognition Workshop (2004)
15. Tu, Z.: Probabilistic boosting-tree: learning discriminative models for classification, recognition, and clustering. In: Tenth IEEE International Conference on Computer Vision, vol. 2, pp. 1589–1596 (2005)
16. Monteith, K., Carroll, J.L., Seppi, K., Martinez, T.: Turning bayesian model averaging into bayesian model combination. In: The 2011 International Joint Conference on Neural Networks, pp. 2657–2663
17. Kuncheva, L.I.: Combining Pattern Classifiers: Methods and Algorithms. Wiley, New Jersey (2004)

18. Yang, P., Yang, Y.H., Zhou, B., Zomaya, A.: A review of ensemble methods in bioinformatics. Current Bioinf. **5**(4), 296–308 (2010)
19. Wu, W., Zhang, S.: Evaluation of error-sensitive attributes. In: Li, J., Cao, L., Wang, C., Tan, K.C., Liu, B., Pei, J., Tseng, V.S. (eds.) PAKDD 2013 Workshops. LNCS, vol. 7867, pp. 283–294. Springer, Heidelberg (2013)
20. Bache, K., Lichman, M.: UCI Machine Learning Repository. University of California, School of Information and Computer Science, Irvine, CA (2013). http://archive.ics.uci.edu/ml
21. Hall, M., Frank, E., Holmes, G., Pfahringer, B., Reutemann, P., Witten, I.H.: The WEKA data mining software: an update. SIGKDD Explor. **11**(1), 10–18 (2009)

Probabilistic Hoeffding Trees

Sped-Up Convergence and Adaption of Online Trees on Changing Data Streams

Jonathan Boidol[1,2]([⊠]), Andreas Hapfelmeier[2],
and Volker Tresp[1,2]

[1] Institute for Computer Science, Ludwig-Maximilians University, Oettingenstr. 67,
80538 München, Germany
boidol@cip.informatik.uni-muenchen.de

[2] Siemens AG, Corporate Technology, Otto-Hahn-Ring 6, 81739 München, Germany
{andreas.hapfelmeier,volker.tresp}@siemens.com

Abstract. Increasingly, data streams are generated from a growing number of small, cheap sensors that monitor, e.g., personal activities, industrial facilities or the natural environment. In these settings, there are often rapid changes in input-to-target relations and we are concerned with tree-structured models that can rapidly adapt to these changes. Based on our new algorithms accuracy and tracking behavior is improved, which we demonstrate for a number of popular tree based-classifiers with over state-of-the-art change detection using five data sets and two different settings. The key novel idea is the representation of record values as distributions rather than point-values in the stream setting, covering a larger part of the instance space early on, and resulting in an often smaller, more flexible classification model.

Keywords: Online decision tree learning · Uncertainty-aware data streams · Classification · Concept change · Regularization

1 Introduction

Recent technological developments have immensely increased the data volume and require new analytic techniques beyond ordinary batch learning. Streaming data analysis is concerned with applications where the records are processed in non stopping streams of information. Examples include the analysis of streams of text, like in twitter, or the analysis of image streams like in flickr or the analysis of video streams. Other applications include large scale remote monitoring of environmental sensors and of industrial sensors where data rates can reach terabytes per day. Also, streams are often be subject to gradual or sudden changes in the relation between attributes (or input) and target variable, and algorithms have to adapt to these changing conditions. A gradual change is termed *concept drift*, a sudden change is called *concept change*. In this work, we will consider the case of concept changes, more specifically changes in the conditional probability $P(y|x)$ of events y given measurements x [14]. Examples are changes caused

© Springer International Publishing Switzerland 2015
P. Perner (Ed.): ICDM 2015, LNAI 9165, pp. 94–108, 2015.
DOI: 10.1007/978-3-319-20910-4_8

by the transition from day to night, by changes in production phases, by the introduction of new features or other sudden changes in the environment.

For these scenarios, online learners for classification have been developed that should meet the following criteria: Learning should operate iteratively, i.e. build a classification model incrementally without needing all the data before training starts. It should use every record in a single pass, i.e. look at every example only once. It should use finite resources, i.e. the algorithm's training time and space requirements should not grow with the data size. It should exhibit any-time readiness, i.e. provide the best possible classification model at any time during execution. Hoeffding Tree-Based classifiers possess most of these desired properties, and remain fairly easy to implement and analyze, and have been shown to be robust and highly scalable. In this work, we aim to improve classification on data streams that undergo concept changes to which the classifier has to react promptly.

The Hoeffding Tree is a classifier that deals with streaming data [7], also known as VFDT (Very Fast Decision Tree), upon which many state-of-the-art Online-Learners build, e.g. FIMTDD [14], CVFDT [13], VFDTc [9], iOVFDT [11], Hoeffding Option Trees [21]. VFDT and its derivatives incrementally build a decision tree and prune parts again as necessary without looking at any record more than once. Splits in the tree are introduced when sufficient examples have been seen to make a confident decision. This decision is guided by statistical bounds, e.g. the eponymous Hoeffding bound, that need only sufficient statistics of fixed size stored in the tree. The nature of these statistics varies but typically allows to calculate the best split on promising attributes. Different pruning criteria have been added to the basic algorithm to detect changes in the underlying data stream and adapt the tree accordingly. If the growth of the tree is suitably checked to avoid unlimited growth – and eventual overfitting –, the whole classifier is therefore in size independent of the size of the data stream.

More recently, methods have been developed that deal with inherent uncertainty in the data that stems e.g. from measurement errors, processing errors, technical limitations or natural fluctuations. The Uncertainty-Aware approach does not assume recorded attribute values as given, but recognizes that attribute values are representative of an underlying probability distribution. Such a situation might also arise if there are multiple measurements, say from redundant sensors in the same environment, without practical means to pick one measurement over the other if they differ. This paper shows how an Uncertainty-Aware handling of the data significantly improves accuracy and any-time-readiness in Online-Classification of changing dynamic streams.

The remainder of the paper is organized as follows. Section 2 reviews related work. Sections 3 and 4 introduce our algorithm in the context of existing decision tree algorithms. Section 5 describes the data sets, we used in our experiments, and shows the success of our algorithm compared to popular Online-Classifiers. Finally, Sect. 6 discusses the conclusions we reached based on these experiments and outlines directions for future research.

2   Related Work

In the last decade, there has been substantial research in the areas of stream processing and uncertain data. Uncertainty-aware research covers topics from clustering uncertain data [6,18,20] or outlier detection [1] to querying probabilistic databases [17]. Classification with tree models has been done by e.g. [25]. The common idea is that the expected distance between two objects is calculated with probability distributions of these objects. This work deals exclusively with static and stationary data but we borrow concepts from the research into uncertainty and apply them to data where uncertainty is not apparent but used as a tool to essentially extract more information from the data.

The earliest stream classification with iterative tree models has been developed by [7] and built upon by [9,11,13,21]. Reference [22] used an uncertainty-aware approach to improve classification models on static data, [19] used a similar approach for online stream-classification. We improve upon their work and extend the analysis to cases with time-changing data streams. To the best of our knowledge, we present the first analysis of an Uncertainty-aware classifier for data streams with concept change.

3   Online Trees for Changing Data

Our algorithm design is based on the basic Hoeffding Tree algorithm, but is in principal adaptable to any tree-like incremental learner. We introduce an elementary notation in the following section and review basic concepts of Online Decision Trees. We then present our approach PHT (Probabilistic Hoeffding Tree) as extension of those trees and outline how these changes can be implemented in an online fashion in the next section.

3.1   Online Decision Trees

Let $A_i = (a_{i,1}, \ldots, a_{i,k})$ be an instance of the data stream with k single-valued attributes where the index i notes the position in the data stream. Like all decision trees, Hoeffding Trees consist of nodes and edges (V, E) where the nodes contain tests to decide which edge to follow towards a leave of the tree. To build a Hoeffding tree, during the training phase leave nodes are recursively replaced with decision nodes. The leave nodes store statistics, decision nodes contain a split attribute. Each instance is assigned to one leaf node v after a series of tests that determine the path from the root down. These tests select the appropriate path based on the relevant split attribute of the instance in each node along the path. Thereby they determine the one branch A_i falls into and the statistics stored in leaf v are updated with the information from A_i. In some versions statistics in the nodes on the path to v are also updated. A decision to grow or prune the tree is then based on these updated statistics. They are also crucial to detect changes in the data stream and adapt the tree via pruning and regrowth

Algorithm 1. Basic Online Tree Induction

Input: data stream s yielding records A_i
Output: decision tree t
1: **procedure** TREEINDUCTION
2:     $t \leftarrow empty\,leaf$
3:     **while** $A_i \leftarrow next\_from(s)$ **do**
4:         $v \leftarrow get\_leaf(A_i)$
5:         $update(v, A_i)$
6:         $test\_and\_split(v)$
7:         $prune(t)$
8:     **end while**
9: **end procedure**

to the changes [2]. The pseudo code to build an incremental tree is given in Algorithm 1.

Note that we will only explicitly consider two-way splits for numeric attributes. More than two branches are possible, and common for categorical attributes, but the case for multi-way splits and discrete distributions follows easily. The tree can at any time be trained further with more instances from the stream and conversely prediction with the induced tree is possible at any point in the lifetime of the tree. Ordinarily, the prediction for a record A_i is based on whatever model is stored in the leaf to which A_i is assigned. In the simplest case this might be a single class-label or numeric value, more sophisticated versions store specific classification or regression models in the leaves.

4   Probabilistic Hoeffding Trees

The main idea in our approach is to treat records not as sets of exactly measured single values but to treat the attributes as a probability density function (PDF) centered around the recorded value instead. We call the resulting class of Hoeffding-tree algorithms PHT (Probabilistic Hoeffding Trees).

4.1   Probabilistic Records for Decision Trees

The modifications compared to the base algorithm are again given as pseudo code in Algorithm 2.

We replace the single value of a_{ij} with a PDF $p(a_{ij})$ centered around a_{ij}. For numeric attributes a uniform or Gaussian distribution are standard choices, for categorical attributes any discrete distribution specified over the possible values of a_{ij} is acceptable [5,23]. The training process is then adapted in the following way: We assume again an initial weight of 1 for every instance A_i. For every test A_i encounters in a node, e.g. $a_{ij} < t_m$, the integrals $w_l = \int_{-\infty}^{t_m} p(a_{ij})\,da_{ij}$ and $w_r = \int_{t_m}^{\infty} p(a_{ij})\,da_{ij}$ for the left and right branch are calculated. w_l and w_r simply determine, how much of the probability mass of the attribute falls in the left and right branch respectively. The values w_l and w_r are then interpreted as the weight of the branch. A_i follows every branch where w is larger than 0 simultaneously and may reach more than one leaf of the tree (cf. line 4 of Algorithm 2). The relative weight of a leaf v is then $w^{A_i,v} = \prod_{m \in M} w_{I,m}$, the product of all weights along the path to leaf v branching at nodes m, where $I \in \{l, r\}$ determines the branch taken at node m. The statistics in these leaves are then updated with the information from A_i, as in the original case, but down-weighted by $w^{A_i,m}$ (cf. line 6 of Algorithm 2). The total weight of v still sums to 1 but it promotes growth in more than a single leaf.

Algorithm 2. Incremental Uncertain Tree Induction

Input: data stream s yielding records A_i
Output: decision tree t
1: **procedure** PROBABILISTICTREEINDUCTION
2:     $t \leftarrow empty\,leaf$
3:     **while** $A_i \leftarrow next\_from(s)$ **do**
4:         $L \leftarrow get\_leaves(A_i)$
5:         **for all** $v \in L$ **do**
6:             $update(v, A_i, rel\_weight(A_i, v))$
7:             $test\_and\_split(v)$
8:         **end for**
9:         $prune(t)$
10:    **end while**
11: **end procedure**

We treat instances for prediction the same way as in training, see the modifications to the prediction process in Algorithm 3. We do not need to change the prediction model used in the tree, but we do not limit the prediction to one of those models. Our algorithm filters one record down to several leaves instead, and averages the predictions from every leaf weighted by $w^{A_i,v}$.

The voting (cf. line 9 in Algorithm 3) has the advantage of giving a distribution for the prediction from which a confidence value can be inferred, even if the base algorithm does not provide one.

In the long run in the stream setting, using a symmetric distribution and using point-values will – assuming a stationary stream – in theory converge. The advantages lie in more independence towards the order of the instances, greater flexibility during training and prediction and – as experiments will show – in the speed of the convergence towards the expected optimal tree.

4.2   Online Approximation of Density Functions

The PDFs for the attribute values have always been chosen as uniform distribution with mean equal to the original attribute point value and a standard deviation proportional to $(b - a) \times w$. Here a and b are the minimum and maximum values for the attribute that actually appear in the data set and w controls

---

Algorithm 3. ProbabilisticTreePrediction

Input: tree t, instance A_i
Output: prediction $\tilde{x}$
1: **procedure** PROBABILISTICTREEPREDICTION
2:     $L \leftarrow get\_leaves(A_i)$
3:     $V = \emptyset$
4:     **for all** $v \in L$ **do**
5:         $vote \leftarrow predict(A_i, v)$
6:         $weight \leftarrow rel\_weight(A_i, v)$
7:         $V = V \cup (vote, weight)$
8:     **end for**
9:     $\tilde{x} \leftarrow average(V)$
10:    **return** $\tilde{x}$
11: **end procedure**

---

the width of the distribution and the 'fuzziness' of the attribute value. For the categorical attributes, the PDF has been constructed in such a way that $1 - w$ of the probability mass is placed onto the original value and the rest spread uniformly on the possible attribute values. For the synthetic data set (with numerical attributes only), a and b have been chosen so that $P(x_a \in [a, b]) \geq 0.997$ or approximately within three standard deviations of the mean.

The notation as range of values is closely related to, but here more intuitive, than the standard deviation. If the attribute range is unknown, it can be estimated from the stream for example with a number of algorithms that incrementally calculate the variance of the attribute, e.g. [16]. The ranges follow easily from the variance, for example for uniform distributions $\sigma^2 = \frac{(b-a)^2}{12}$.

Representing the PDF $p(a_{ij})$ is simple if the attribute j is categorical. Then we need only the probability for every possible value of j which has a finite and in practice usually small domain. In principal, numeric attributes could be discretized in a number of bins and treated equivalently [19]. This, however, discards the ordinality of the attribute values, forces multi-way splits and is necessarily low grained. In a simple, non-analytical solution, which has been used for example in [25], the PDF can be represented numerically by storing a set of s sample points drawn from $p(a_{ij})$ which approximates any function with a discrete distribution. Conveniently, this works equally well for numeric and categorical attributes and for all types of distributions. We chose $s = 100$ which provided a balance between approximation quality and performance in our tests.

5   Experiments

To test our algorithm we used 4 large data sets collected from sensor readings or network streams and one synthetic data set. The real data sets are all available at the UCI machine learning repository and range from 5 k to 580 k in size. While these are sufficient to gauge the algorithm behavior, we also use synthetic data to test performance in longer runs. For those experiments we used instance streams of 5 million instances. All test runs have been performed on a PC with an Intel Xeon 1.80 GHz CPU, running Linux with a 2.6.32 x86_64 kernel, and with memory limitations set to 64 MB.

5.1   Implementation

We adapted three different tree induction algorithm to PHT: Adaptive Hoeffding Trees [2], iOVFDT [11] and Hoeffding Option Trees [21] to HoeffdingAdaptive-TreePHT, iOVFDTPHT and HoeffdingOptionTreePHT. The implementation was done in the MOA framework [3], where reference implementations of the aforementioned algorithms exist and the algorithms could easily be extended.

For the evaluation of the experiments, we recorded accuracy, resulting tree size and training time measured in an interleaved test-then-train setting where every instance is first used for blind testing, and then to train the tree [2]. The standard deviation for each measure is computed over 10 repeated experiments with shuffled data sets or different initialization parameters for the synthetic data set. For the accuracy we use a fading average as described in [10] with a fading factor α of 0.99. The fading average $M_\alpha(i)$ is defined as

$$M_\alpha(i) = \frac{S_\alpha(i)}{N_\alpha(i)} \tag{1}$$

$$S_\alpha(i) = I_i + \alpha \times S_\alpha(i-1); \ S_\alpha(1) = I_1 \tag{2}$$

$$N_\alpha(i) = 1 + \alpha \times N_\alpha(i-1); \ N_\alpha(1) = 1, \tag{3}$$

where S_α is the fading sum of observations, N_α the fading increment and $I = 1$ for a correct prediction, 0 otherwise.

We report tree size in number of nodes rather than model size in bytes. The consumed memory depends not only on the implementation but also on the number and types of attributes in a data set. The number of nodes on the other hand allows an easier comparison of different tree models. We test our algorithms first on the static data sets to establish their performance and advance to time changing data streams in the following sections.

5.2   Data Sets

Robot Movement Data (RM). The RM data set is available since 2010. It contains 24 numeric attributes recorded from the a robot's sensors and four distinct classes, which determine the robot's course along a wall. The data set contains 5,456 instances [8].

Person Activity Analysis (PA). The PA data set is available since 2010. It recorded the instances collected from four sensors placed on both ankles, belt and chest of five people. Each instance has five numeric attributes, two categorical attributes and one of eleven classes. The classes distinguish human activities, e.g. walking, standing, falling, etc. The data set contains 164,860 instances [15].

Network Attack Detection (NA). The NA data set has been published for the KDD CUP 1999. It describes network connections and is used to classify normal and abnormal connections, i.e. attacks . It contains 34 numeric and seven categorical attributes like duration, error rate and protocol type. The connection types are distinguished in 23 distinct classes. We use 10 % of the full data with 494,021 instances [24].

Cover Type (CT). The CT data set is available since 1999. It collects surveillance sensor data of forestland. Each instance provides 42 categorical attributes and eleven numeric attributes like soil type, elevation, and hill shade. It distinguishes cover types in seven classes. The data set contains 581,012 instances [4].

Synthetic RBF Stream (RBF). This type of synthetic stream uses a radial basis function to generate arbitrarily large data sets. Using different initialization parameters we can create different streams, each of arbitrary length. The streams for the experiments were initiated with fixed seeds to ensure reproducibility. We set the parameters to use 50 base functions that generate 15 attributes and 4 classes and limited stream size to five million instances.

5.3   Results on Static Data

We implemented as PHT variants the following classifiers: HoeffdingAdaptive PHT, iOVFDTPHT and HoeffdingOptionPHT. We tested these on the five large data sets described in Sect. 5.2 and varied the values for the width w of the assumed distribution from 0 to 0.5. $w = 0$ means no uncertainty and is equivalent to the base classifiers our algorithms build upon. In general, we see an improvement for $w \leq 0.1$, with small to moderate (3.3 %) improvement of accuracy. Accuracy drops for larger values of w that would imply major uncertainty and are not reported.

Taking the best performing setting for each classifier and data set, we see an improvement in 10 out if 15 cases (each significant with $p < 0.1$, in a one sided t-test) in Table 1. Figure 1 shows the final accuracy for the smallest (RM) and the largest (CT) UCI data set. The fading accuracy used gives less weight to the earlier test examples, giving an overall accuracy that favors the recent predictions.

Figure 2 shows the accuracy during the lifetime of the data stream of the RBF data set. We used the RBF data set to analyze the behavior of the algorithms on much longer lived data streams and see improvement over the base classifiers, especially for the Hoeffding Adaptive Tree.

The tree size on average stays within two nodes of the base classifier, with a few exceptions on the larger data sets. There is no clear correlation between changes in model size and improved performance, with five of the eight improved models being smaller, three larger than the base classifier. Tree size does, on the other hand, decrease slightly with increasing values of w since a flatter distribution makes splits in the tree less likely.

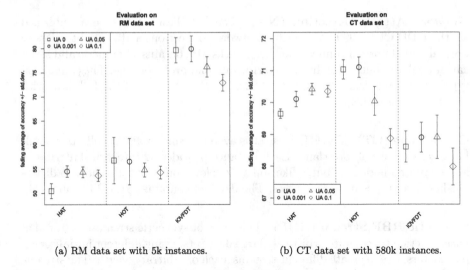

(a) RM data set with 5k instances.   (b) CT data set with 580k instances.

Fig. 1. Comparison of different flavors of Hoeffding-Tree based classifiers (iOVFDT-PHT (iOVFDTPHT), HoeffdingOptionTreePHT (HOTPHT), HoeffdingAdaptiveTree (HATPHT)) on large UCI data sets. Shown here are only the smallest and the largest of the used data sets. UA gives the width-parameter of the PDF replacing the attribute values. Accuracy is the accuracy with a fading factor of 99%. The standard deviation is calculated from 10 shuffled runs.

The most interesting observation here is, how even very small values for w can improve the classification without major cost to the model. Running time for the best performing models with $w \neq 0$ stays within a factor of 2 to the run time of the base classifier. Our algorithms (with $w \leq 10\,\%$) hold equal to or considerably outperform the base algorithms.

5.4   Significant Improvements During Concept Change

While stationary streams are much easier to deal with, we expect both gradual and sudden changes in real life streams. We therefore especially studied the effects of concept change, i.e. changes in the conditional probability of the classes given attribute vectors, and the improvements our algorithm achieves in such a setting. This occurs if an observed stream/the underlying system undergoes different phases in its lifetime like seasonal changes, day-night cycles in ecological systems, or different production phases in industrial machinery. Normal behavior might look completely different before and after these changes and the classification algorithm has to adapt accordingly. While HoeffdingAdaptiveTrees and HoeffdingOptionTrees have the capability to detect changes and adapt, iOVFDT does not have a mechanism to adapt to dynamic streams and is not included in this section.

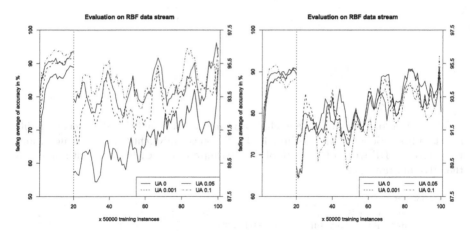

(a) HoeffdingAdaptiveTreePHT on RBF data stream with 5M instances.

(b) HoeffdingOptionTreePHT on RBF data stream with 5M instances.

Fig. 2. Evaluation on RBF stream. The plots shows the accuracy over a the lifetime of a stream with 5 million instances exemplary for the Hoeffding Adaptive Trees and Hoeffding Option Trees. The vertical axis is enlarged from the 20%-mark on.

To evaluate the effect of changes for the other classifiers, we streamed each of the original UCI data sets in the stream five times in a row, but at each repetition switched the order of the numeric attributes as suggested by e.g. [12,26,27]. This permutation of attributes induces the desired concept changes and at the same time keeps the integrity of the data set compared to, say, introducing bias or noise into the data. In the RBF data set we simulated such changes by changing the parameters of the generating function. Figure 3 shows the classification accuracy during the lifetime of a RBF-stream with concept changes every 200.000 records. To compare the accuracy over the total lifetime, we averaged the accuracy over 100 sample points during the stream life time and report the results in Table 2. After every concept change, all classifiers fall back in accuracy and recover gradually, but our algorithms recovers at a much faster rate and in this setting of changing streams significantly ($p < 0.1$ in a one-sided t-test for the best-performing setting) outperforms the base classifiers. Figure 3 show how accuracy behaves before and after the concept change. For the smaller data set shown in Fig. 3(c) the break-down between change is less pronounced since the classifier has not reached a stable plateau before the concept change as in the longer-lived streams. Our algorithm improves the results of the base classifiers by up to 16% with an average improvement of 3.2%. We improve over HoeffdingOptionTree in 3 out of 5 data sets and over HoeffdingAdaptiveTree in 5 out of five data sets with no significant increase in model size.

Table 1. Tree size of final tree and classification accuracy on 4 UCI data sets with different width-options for the attribute PDFs. UA00 is equivalent to the base algorithm, UAmin to a PDF with 0.1 % width of the attribute range, UA05 to a width of 5 % of the attribute range, etc.

RM data set						
	Accuracy in %			Tree size in #nodes		
	HOTPHT	HATPHT	iOVFDTPHT	HOTPHT	HATPHT	iOVFDTPHT
UA00	**56.81** ± 4.83	50.49 ± 1.54	79.68 ± 2.61	0.60 ± 1.26	0.00 ± 0.00	5.90 ± 0.57
UAmin	56.58 ± 4.81	**54.54** ± 1.18	**79.95** ± 2.67	0.60 ± 1.26	0.00 ± 0.00	6.10 ± 0.74
UA05	54.84 ± 1.45	54.38 ± 1.11	76.21 ± 2.12	0.20 ± 0.63	0.00 ± 0.00	6.90 ± 0.88
UA10	54.30 ± 1.30	53.67 ±1.19	72.90 ± 1.73	0.40 ± 0.84	0.00 ± 0.00	6.80 ± 0.63
PA data set						
	Accuracy in %			Tree size in #nodes		
UA00	50.12 ± 0.29	46.84 ± 0.15	**39.43** ± 0.22	3.30 ± 0.48	4.00 ± 0.00	3.80 ± 0.42
UAmin	**50.21** ± 0.28	**48.10** ± 0.44	39.35 ± 0.34	3.20 ± 0.42	3.20 ± 0.42	3.90 ± 0.32
UA05	49.80 ± 0.36	47.90 ± 0.53	39.22 ± 0.31	3.10 ± 0.32	3.40 ± 0.52	3.90 ± 0.32
UA10	48.75 ± 0.21	47.19 ± 0.66	38.40 ± 0.43	3.00 ± 0.00	3.10 ± 0.57	3.70 ± 0.48
NA data set						
	Accuracy in %			Tree size in #nodes		
UA00	**99.70** ± 0.04	98.34 ± 0.02	**98.89** ± 0.17	4.60 ± 1.26	2.10 ± 0.32	3.80 ± 0.92
UAmin	99.38 ± 0.17	99.08 ± 0.23	98.79 ± 0.15	3.00 ± 0.00	3.70 ± 0.48	4.30 ± 0.67
UA05	99.35 ± 0.17	**99.11** ± 0.20	98.84 ± 0.15	3.20 ± 0.42	4.20 ± 0.42	3.60 ± 0.70
UA10	99.04 ± 0.15	98.75 ± 0.38	98.41 ± 0.22	6.50 ± 1.18	3.70 ± 0.67	5.20 ± 0.42
CT data set						
	Accuracy in %			Tree size in #nodes		
UA00	71.04 ± 0.30	69.65 ± 0.15	68.61 ± 0.49	10.20 ± 1.40	8.33 ± 1.12	6.30 ± 1.16
UAmin	**71.11** ± 0.32	70.11 ± 0.24	68.90 ± 0.47	10.00 ± 1.70	13.00 ± 1.94	7.30 ± 0.82
UA05	70.04 ± 0.45	**70.42** ± 0.18	**68.91** ± 0.68	8.90 ± 0.74	9.56 ± 0.73	7.60 ± 1.07
UA10	68.87 ± 0.30	70.35 ± 0.18	67.98 ± 0.58	7.80 ± 0.63	9.78 ± 1.39	7.60 ± 0.84
RBF data set						
	Accuracy in %			Tree size in #nodes		
UA00	91.73 ± 1.21	91.72 ± 0.04	**88.68** ± 5.21	22.00 ± 7.07	27.00 ± 0.00	9.50 ± 0.71
UAmin	92.14 ± 1.70	**93.33** ± 0.02	83.75 ± 4.55	22.00 ± 2.83	26.00 ± 0.00	10.00 ± 0.00
UA05	**92.89** ± 1.49	92.52 ± 0.03	84.29 ± 0.42	23.50 ± 12.02	32.00 ± 0.00	9.50 ± 0.71
UA10	92.45 ± 1.86	90.85 ± 0.07	79.62 ± 0.08	21.00 ± 7.07	26.00 ± 0.00	8.50 ± 0.71

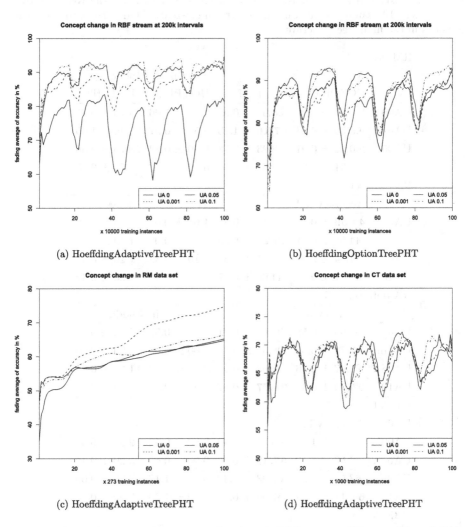

Fig. 3. Effect of concept changes simulated with RBF stream (Figures (a) and (b)) and UCI data sets (figures (c) and (d)). UA gives the width-parameter of the PDF replacing the attribute values. Accuracy is the accuracy with a fading factor of 99 %.

Table 2. Accuracy and model size on the data sets with 4 induced concept changes. UA00 is equivalent to the base algorithm, UAmin to a PDF 0.1 % width of the attribute range, UA05 to a width of 5 % of the attribute range, etc. Accuracy is the average of the accuracies at 100 sample points during the stream existence. Model size is the number of nodes in the final tree structure.

RM data set × 5

	Accuracy in %		Tree size in #nodes	
	HOTPHT	HATPHT	HOTPHT	HATPHT
UA00	**65.43 ± 4.07**	58.65 ± 0.97	8.30 ± 1.06	5.70 ± 0.67
UAmin	65.09 ± 4.23	**67.16 ± 1.73**	8.50 ± 0.97	5.40 ± 1.65
UA05	60.42 ± 0.91	62.23 ± 3.78	6.70 ± 0.82	8.70 ± 2.58
UA10	58.31 ± 0.61	63.62 ± 3.55	6.70 ± 0.67	6.50 ± 2.55

PA data set × 5

	Accuracy in %		Tree size in #nodes	
UA00	**44.66 ± 0.89**	43.11 ± 0.48	1.00 ± 0.00	1.10 ± 0.32
UAmin	44.66 ± 0.91	**45.22 ± 0.33**	1.00 ± 0.00	1.40 ± 0.52
UA05	44.34 ± 0.94	45.05 ± 0.32	1.10 ± 0.32	1.20 ± 0.42
UA10	43.81 ± 0.94	44.60 ± 0.27	1.10 ± 0.32	1.20 ± 0.42

NA data set × 5

	Accuracy in %		Tree size in #nodes	
UA00	97.65 ± 0.58	97.61 ± 0.21	3.90 ± 0.57	2.70 ± 0.48
UAmin	97.72 ± 0.56	98.31 ± 0.51	3.00 ± 0.00	1.00 ± 0.00
UA05	97.97 ± 0.46	98.63 ± 0.28	3.60 ± 0.52	1.00 ± 0.00
UA10	**97.99 ± 0.45**	**98.77 ± 0.27**	3.40 ± 0.52	1.00 ± 0.00

CT data set × 5

	Accuracy in %		Tree size in #nodes	
UA00	63.81 ± 1.12	65.81 ± 0.50	8.90 ± 0.57	2.50 ± 0.53
UAmin	63.67 ± 1.10	**67.45 ± 0.46**	9.00 ± 0.67	3.20 ± 0.42
UA05	63.59 ± 0.95	67.23 ± 0.32	8.60 ± 1.17	3.00 ± 0.00
UA10	**64.41 ± 1.04**	67.05 ± 0.52	9.10 ± 1.20	2.90 ± 0.32

RBF data set × 5

	Accuracy in %		Tree size in #nodes	
UA00	84.70 ± 1.16	73.52 ± 2.40	12.50 ± 0.58	9.40 ± 2.70
UAmin	85.62 ± 0.69	85.99 ± 0.37	13.25 ± 1.26	9.80 ± 0.84
UA05	87.15 ± 0.41	88.61 ± 0.70	13.50 ± 1.00	10.00 ± 0.00
UA10	**87.18 ± 0.98**	**89.71 ± 0.70**	13.00 ± 0.82	10.20 ± 0.45

6   Conclusion and Further Work

In this paper we have shown a generic approach that extends stream classification models to incorporate the concept of uncertain data. We tested this approach on several classifiers and data sets and achieved significantly improved accuracy with comparable model size and run time across all data sets and classifiers we examined. In the case of data sets with concept change we improve accuracy by up to 16 % with 3.2 % on average. Our approach reacts swiftly to changing data streams which makes it especially suited to environments where the concept generating the streams changes periodically, as is the case in many industrial or ecological applications. Non-synthetic data where the data quality is quantified, i.e. the actual uncertainty of measured values is known appears not to be available at the moment. If such data sets become accessible, we expect much interesting results if we could substitute empirical values for the idealized PDFs. Also, we believe that the approach of uncertainty-aware data can be broadened to other types of algorithms, not limited to classification or tree-like prediction models.

References

1. Aggarwal, C.C., Philip, S.Y.: Outlier detection with uncertain data. In: SDM, pp. 483–493. SIAM (2008)
2. Bifet, A., Gavaldà, R.: Adaptive learning from evolving data streams. In: Adams, N.M., Robardet, C., Siebes, A., Boulicaut, J.-F. (eds.) IDA 2009. LNCS, vol. 5772, pp. 249–260. Springer, Heidelberg (2009)
3. Bifet, A., Holmes, G., Kirkby, R., et al.: Moa: Massive online analysis. J. Mach. Learn. Res. **11**, 1601–1604 (2010)
4. Blackard, J.A., Dean, D.J.: Comparative accuracies of artificial neural networks and discriminant analysis in predicting forest cover types from cartographic variables. Comput. Electron. Agric. **24**(3), 131–151 (1999)
5. Cheng, R., Kalashnikov, D.V., Prabhakar, S.: Evaluating probabilistic queries over imprecise data. In: Proceedings of the 2003 ACM SIGMOD International Conference on Management of Data, pp. 551–562. ACM (2003)
6. Cormode, G., McGregor, A.: Approximation algorithms for clustering uncertain data. In: Proceedings of the Twenty-Seventh ACM SIGMOD-SIGACT-SIGART Symposium on Principles of Database Systems, pp. 191–200. ACM (2008)
7. Domingos, P., Hulten, G.: Mining high-speed data streams. In: Proceedings of the Sixth ACM SIGKDD International Conference on Knowledge Discovery and Data Mining, pp. 71–80. ACM (2000)
8. Freire, A.L., Barreto, G.A., Veloso, M., et al.: Short-term memory mechanisms in neural network learning of robot navigation tasks: A case study. In: 2009 6th Latin American Robotics Symposium (LARS), pp. 1–6. IEEE (2009)
9. Gama, J., Medas, P., Rodrigues, P.: Learning decision trees from dynamic data streams. In: Proceedings of the 2005 ACM Symposium on Applied Computing, pp. 573–577. ACM (2005)
10. Gama, J., Sebastião, R., Rodrigues, P.P.: On evaluating stream learning algorithms. Mach. Learn. **90**(3), 317–346 (2013)

11. Hang, Y., Fong, S.: Stream mining dynamic data by using iOVFDT. J. Emerg. Technol. Web Intell. **5**(1), 78–86 (2013)

12. Hashemi, S., Yang, Y., Mirzamomen, Z., et al.: Adapted one-versus-all decision trees for data stream classification. IEEE Trans. Knowl. Data Eng. **21**(5), 624–637 (2009)

13. Hulten, G., Spencer, L., Domingos, P.: Mining time-changing data streams. In: Proceedings of the Seventh ACM SIGKDD International Conference on Knowledge Discovery and Data Mining, pp. 97–106. ACM (2001)

14. Ikonomovska, E., Gama, J.: Learning model trees from data streams. In: Boulicaut, J.-F., Berthold, M.R., Horváth, T. (eds.) DS 2008. LNCS (LNAI), vol. 5255, pp. 52–63. Springer, Heidelberg (2008)

15. Kaluža, B., Mirchevska, V., Dovgan, E., Luštrek, M., Gams, M.: An agent-based approach to care in independent living. In: de Ruyter, B., Wichert, R., Keyson, D.V., Markopoulos, P., Streitz, N., Divitini, M., Georgantas, N., Mana Gomez, A. (eds.) AmI 2010. LNCS, vol. 6439, pp. 177–186. Springer, Heidelberg (2010)

16. Knuth, D.E.: The Art of Computer Programming. Seminumerical Algorithms, 3rd edn., vol. 2, p. 232. Addison-Wesley, Boston (1998)

17. Kriegel, H.P., Bernecker, T., Renz, M., et al.: Probabilistic Join Queries in Uncertain Databases (A Survey of Join Methods for uncertain data), vol. 35. Springer (2010)

18. Kriegel, H.P., Pfeifle, M.: Density-based clustering of uncertain data. In: Proceedings of the Eleventh ACM SIGKDD International Conference on Knowledge Discovery in Data Mining, pp. 672–677. ACM (2005)

19. Liang, C., Zhang, Y., Song, Q.: Decision tree for dynamic and uncertain data streams. In: ACML, pp. 209–224 (2010)

20. Ngai, W.K., Kao, B., Chui, C.K., et al.: Efficient clustering of uncertain data. In: Sixth International Conference on Data Mining, ICDM 2006, pp. 436–445. IEEE (2006)

21. Pfahringer, B., Holmes, G., Kirkby, R.: New options for hoeffding trees. In: Orgun, M.A., Thornton, J. (eds.) AI 2007. LNCS (LNAI), vol. 4830, pp. 90–99. Springer, Heidelberg (2007)

22. Qin, B., Xia, Y., Li, F.: DTU: a decision tree for uncertain data. In: Theeramunkong, T., Kijsirikul, B., Cercone, N., Ho, T.-B. (eds.) PAKDD 2009. LNCS, vol. 5476, pp. 4–15. Springer, Heidelberg (2009)

23. Singh, S., Mayfield, C., Prabhakar, S., et al.: Indexing uncertain categorical data. In: IEEE 23rd International Conference on Data Engineering, ICDE 2007, pp. 616–625. IEEE (2007)

24. Stolfo, S.J., Fan, W., Lee, W., et al.: Cost-based modeling for fraud and intrusion detection: Results from the JAM project. In: Proceedings of the DARPA Information Survivability Conference and Exposition, DISCEX 2000, vol. 2, pp. 130–144. IEEE (2000)

25. Tsang, S., Kao, B., Yip, K.Y., et al.: Decision trees for uncertain data. IEEE Trans. Knowl. Data Eng. **23**(1), 64–78 (2011)

26. Wang, P., Wang, H., Wu, X., et al.: On reducing classifier granularity in mining concept-drifting data streams. In: Fifth IEEE International Conference on Data Mining, 8-p. IEEE (2005)

27. Yang, Y., Wu, X., Zhu, X.: Combining proactive and reactive predictions for data streams. In: Proceedings of the Eleventh ACM SIGKDD International Conference on Knowledge Discovery in Data Mining, pp. 710–715. ACM (2005)

Fast and Robust Supervised Learning in High Dimensions Using the Geometry of the Data

Ujjal Kumar Mukherjee, Subhabrata Majumdar, and Snigdhansu Chatterjee[⊠]

University of Minnesota, Minneapolis, MN 55455, USA
chatt019@umn.edu

Abstract. We develop a method for tracing out the shape of a cloud of sample observations, in arbitrary dimensions, called the *data cloud wrapper* (DCW). The DCW have strong theoretical properties, have algorithmic scalability and parallel computational features. We further use the DCW to develop a new fast, robust and accurate classification method in high dimensions, called the **geometric learning algorithm** (GLA). Two of the main features of the proposed algorithm are that there are no assumptions made about the geometric properties of the underlying data generating distribution, and that there are no parametric or other restrictive assumptions made either for the data or the algorithm. The proposed methods are typically faster and more robust than established classification techniques, while being comparably accurate in most cases.

1 Introduction

We propose a new method for classification, that respects the inherent geometry of the data cloud for each labeled group of observations, and this method is not subject to curse of dimensionality. Our method is based on *multivariate quantiles*, which generalize the notion of quantiles for observations in dimensions greater than one. Arising naturally from the concept of multivariate quantiles is the notion of *data depth*, which is a relative measure of proximity of a given point in space to a collection of observations. For any new or unlabeled observation in the feature space, we estimate the label by computing its depth from the data clouds corresponding to a training data.

There are two important properties of the classification technique presented below. First, we do not make assumptions about the geometrical features of the multidimensional data (for which we use the term *data cloud*) corresponding to the various labels in the training sample. Thus, the proposed method respects the geometric properties of the data, and does not impose shape restrictions on it, hence we call it *geometric learning* algorithm (GLA hereafter). Second, our method is scalable and parallelizable with respect to dimensions and sample size, and does not suffer from the curse of dimensionality, and hence is extremely fast in implementation. It can be seen from the development below that our proposed method extends readily to several other supervised learning problems.

The strengths of the proposed geometric learning method arises from the fact that it is based on multivariate quantiles. In Sect. 2 we discuss these quantiles in

© Springer International Publishing Switzerland 2015
P. Perner (Ed.): ICDM 2015, LNAI 9165, pp. 109–123, 2015.
DOI: 10.1007/978-3-319-20910-4_9

details. Based on projection quantiles, which are a form of multivariate quantiles, we develop a *Data Cloud Wrap* (DCW) procedure that provides a very accurate description of the geometry of any sized data set in any dimension. As an illustration, consider Fig. 1, which contains two bivariate scatter plots, the left panel being that of observations from a Gaussian distribution and the right panel is where observations are from a mixture of two Gaussian distributions. The red curves are obtained by the DCW procedure, and it can be seen that these curves quite accurately capture the geometry of the layout of the observations in either panel. The blue curves in either panel correspond to a *projection quantile* (PQ), which reasonably trace the shapes of the data clouds, but not as accurately as the DCW curves. The black curves are obtained by presuming a Gaussian distribution for the data, with only mean and functions as unknowns. Notice that while this is adequate for capturing the shape of the data cloud when the Gaussian assumption holds, it is a severe misfit when the assumption is violated. The regions enclosed by the different curves in either panel are not expected to have identical probabilistic coverage, owing to different mathematical properties.

Note that in high dimensions, it is essentially impossible to graphically or otherwise elicit how and where assumptions like Gaussian shape of the data geometry are violated. Even if such elicitation were feasible, it is unclear how to use that information for supervised learning, or other data-related tasks. The *geometric learning algorithm* we present here is a clear alternative, that does not rely on such encumbering assumptions.

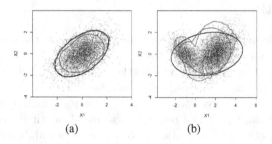

(a)   (b)

Fig. 1. Comparison of usual projection quantiles (blue) with weighted projection quantiles (red), along with a Gaussian confidence ellipsoid (black) for a Gaussian scatter in (a) and mixture of Gaussians in (b). Areas under the different curves are not expected to be equal (Color figure online).

We discuss below how the sets of Fig. 1, and their enclosing boundary curves, may be indexed by vectors of the unit sphere in the feature space. Curves and enclosed sets as in Fig. 1 are fast and accurate visualization tools that are easily available from the proposed procedure. These graphical techniques are naturally best suited for two and three dimensional projections of the data, however, the construction of such sets in any dimension is simple and quick in our proposed methodology. We can take full advantage of distributed and parallel computing tools for this purpose, since the constructions of sets like the ones depicted in

Fig. 1 is linear in both dimension (p) of the feature space, and the number of observations (n), and parallelizable in both dimensions and sample size. Since the DCW is central to the geometric learning procedure, we present theoretical properties of it in Sect. 3.

We also use the DCW algorithm to compute the *data-depth* of any point in the feature space, with respect to any probability distribution function, or data cloud. A data-depth is a relative measure of how close is the given point in space to the center of a data cloud or a median of a (multivariate) probability distribution function. We discuss technical results of data-depths in the current context in Sect. 4 below. The crucial component of obtaining the depth of a given point with respect to a cloud of observations is to project the observations *in a single direction*, which is extremely fast and easy, apart from being an simple parallel procedure.

One immediate application of the DCW and the related data-depth algorithm is in supervised learning, presented in Sect. 5. Owing to the speed and efficiency of the DCW and data-depth algorithms, such classification of observations can be carried out extremely quickly, and the proposed **geometric learning** procedure may be used for *online supervised learning*. Thus, this can be adapted for a real-time analytics tool for classification in big data. Moreover, since we do not make assumptions about the geometry of the description of the data cloud $\mathbf{X}_k$ for any k, the proposed procedure is *robust* against failures of statistical assumption. Note that most statistical assumptions are essentially unverifiable declarations in high-dimensional data, hence such robustness properties are essential.

Apart from being fast and robust, the geometric learning procedure is surprisingly versatile and efficient. In the different datasets we have analysed, some of which are presented below, it seems that the proposed procedure is competitive, if not better, than standard supervised learning methods that are in popular usage. Note however, our goal here is to obtain (i) the shape of the data cloud, and (ii) fast classification without encumbering assumptions, and we do not claim to have an algorithm that will be "most accurate always". It nevertheless turns out that the proposed methodology that is typically hundreds or thousands of times faster than, say, random forest or support vector machine-based algorithm, we generally have a comparably high classification accuracy.

Results are presented in Sect. 7 for some simulated data examples, and in Sect. 8 for several real data examples. We conclude this paper with Sect. 9, where we present some caveats about using geometric learning, and some future research directions.

2   The Projection Quantile

We denote the open unit ball in p-dimensional Euclidean plane as $\mathcal{B}_p = \{x \in \mathbb{R}^p : ||x|| < 1\}$. The notation $||\mathbf{a}||$ stands for the Euclidean norm of a vector $\mathbf{a}$, while $\langle \mathbf{a}, \mathbf{b} \rangle$ stands for the Euclidean inner product between two vectors. For convenience, we reserve the notation $\mathbf{0}$ for a vector of zeroes, and $\mathbf{1}$ for a vector of ones, in appropriate dimensions that will be specified in the right contexts.

Also, we reserve the notation **u** to denote a typical element in this open unit ball. We further reserve the notation $\mathbf{e_u}$ for the unit vector in the direction of $\mathbf{u} \in \mathcal{B}^p$. Thus, $\mathbf{e_u} = \frac{\mathbf{u}}{||\mathbf{u}||}$ when $\mathbf{u} \neq \mathbf{0} \in \mathbb{R}^p$ and $\mathbf{0}$ otherwise. For any vector $x \in \mathbb{R}^p$, we define $x_\mathbf{u} = \langle x, \mathbf{e_u} \rangle$. The projection of x in the direction of $\mathbf{u}$ is, $x_\mathbf{u}\mathbf{e_u} = ||\mathbf{u}||^{-2}\langle \mathbf{x}, \mathbf{u} \rangle \mathbf{u}$.

Let $X \in \mathbb{R}^p$ be a random variable in p-dimensional Euclidean space. For the moment, assume that the center of the distribution of X is the origin. Let, $\mathbf{q_u}$ be the $(1 + ||\mathbf{u}||)/2$-th quantile of $X_\mathbf{u}$, that is, $\mathbb{P}[X_\mathbf{u} \leq \mathbf{q_u}] = (1 + ||\mathbf{u}||)/2$. The **u**-th projection quantile (PQ) is defined in [9] as

$$Q_{proj}(\mathbf{u}) = \mathbf{q_u}\frac{\mathbf{u}}{||\mathbf{u}||} = \mathbf{q_u}\mathbf{e_u}. \tag{1}$$

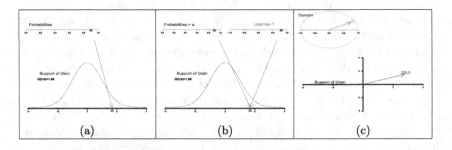

Fig. 2. A graphical depiction of the quantile function in one and two dimensions

The PQ (projection quantile) has several interesting properties, which makes it attractive from both theoretical and algorithmic points of view. It is linearly dependent on the number of dimensions p in calculation of $X_\mathbf{u}$. The sample PQ computation is linear in n also. Additionally, it can be easily seen that the computation of projection quantiles in different directions are unrelated to each other, and can be trivially distributed over a network of computing cores.

We present a brief motivation of the above PQ here, by using the illustrative example of univariate and bivariate Gaussian random variables. Note that for a real random variable X, the quantile function is defined on the interval $[0, 1]$ of probabilities and has as its range as the support of the random variable, and is traditionally defined as $Q(a) = \inf\{q : \mathbb{P}[X \leq q] \geq a\}$ for any $a \in [0, 1]$. For the standard Gaussian distribution, this is illustrated in the left panel of Fig. 2. Note, however, the following is also true [3,4]:

Theorem 1. *The ath quantile is the smallest minimizer of the function* $\mathbb{E}$ $[|X - q| + (2a - 1)(X - q)]$.

Existence and uniqueness of $Q(a)$ is not an issue, owing to convexity of the criterion function, and hereafter we assume adequate conditions to ensure that in the *population*, the above convex criterion function has a unique minimizer. Assuming that the random variable X is absolutely continuous is sufficient for this purpose, and hereafter we assume all feature vectors are absolutely continuous random variables.

In view of the above, we may alternatively define the quantile function as being indexed by $u = 2a - 1 \in [-1, 1]$, and $Q(u)$ as the (unique) minimizer of the convex function $\mathbb{E}\left[\|X - q\| + u(X - q)\right]$, as illustrated by the middle panel in Fig. 2. This definition of a quantile function was extended by [3] for p-dimensional random variables as being indexed by vectors $\mathbf{u} \in \mathcal{B}_p = \{x \in \mathbb{R}^p : \|x\| < 1\}$, and defined as minimizers $Q(\mathbf{u})$ of $\mathbb{E}\left[\|X - q\| + <\mathbf{u}, X - q>\right]$, as illustrated in the right panel of Fig. 2. This is a generalization of one of the earliest attempts at defining multivariate median by [7]. Note that Chaudhuri's multivariate quantiles cannot be computed for $p > n$ using the algorithm given in [3], and requires iterative methods for even $p \leq n$. Additionally, it was seem that (a) this definition of multivariate quantiles does not capture the data geometry adequately, and (b) $Q(\mathbf{u})$ and $\mathbf{u}$ were nearly parallel in several simulated data examples. These observations motivate the projection quantile, where instead of using the full Euclidean norm of $X - q$, we only use that part of $X - q$ that is parallel to $\mathbf{u}$. Some amount of algebra reduces this procedure to the description of PQ provided above, and Fig. 1 shows its efficacy.

3   The Data Cloud Wrapper

The PQ described above does not fully capture the shape of the data geometry, mainly because of two issues. First, the spread of the data in different directions $\mathbf{e_u}$ from the center is different, and PQ does not accomodate for that. Second, all information related to any feature vector X_i in the directions orthogonal to $\mathbf{e_u}$ is discarded. The *data cloud wrapper* (DCW) algorithm attempts to correct these two discrepancies in the PQ, by introducing two weight functions. First, we adjust the *direction specific scaling* using $w_\mathbf{u}$ defined below. Then, we incorporate the information from the ith observation in the directions orthogonal to $\mathbf{e_u}$ using another weight factor w_{2i}, also detailed below.

Recall that in accordance with the notation developed earlier, $X_{\mathbf{u}i}\mathbf{e_u}$ is the projection of X_i along $\mathbf{e_u}$. After centering (at the co-ordinatewise median) and scaling (using the median absolute deviation) the data, we first compute $Q_{proj}(\mathbf{u})$, the projection quantile along $\mathbf{u}$. We then compute global weights for the direction vector $\mathbf{u}$ by k-mean distance. Define d_i is the Euclidean distance of X_i from $Q_{proj}(\mathbf{u})$, and $d_{(1)} < \ldots < d_{(n)}$ are the ordered distances. We then define the k-mean distance as $\bar{d}_k = \frac{1}{n} \sum_{i=1}^{n} d_i \mathbb{I}_{\{d_i < d_{(k)}\}}$. Here, k is a tuning parameter that we choose depending on the application. We then define $w_\mathbf{u} = \exp(-a\bar{d}_k)$ as a scaling factor to be used in the direction $\mathbf{e_u}$. Our next step is to compute the norms of the vectors $\|X_{\mathbf{u}\perp i}\| = \|X_i - X_{\mathbf{u}i}\mathbf{e_u}\|$, which we use in the weight function $w_{2i} = \exp\left[-b\frac{\|X_{\mathbf{u}\perp i}\|}{\|X_i\|}\right]\mathbb{I}_{\{\|X_{\mathbf{u}\perp i}\| \leq \epsilon\}}$. Here, b and ϵ are tuning parameters. Suppose $\{j_1, \ldots, j_{n_u}\}$ are the indices for which w_{2i} is non-zero. We now define $\tilde{X}_{\mathbf{u}j_k} = w_\mathbf{u} w_{2j_k} X_{\mathbf{u}j_k}$, for $k = 1, \ldots, n_u$. The DCW in the direction $\mathbf{e_u}$ is obtained by selecting the $\alpha = (1 + \|\mathbf{u}\|)/2$-th quantile of $\tilde{X}_{\mathbf{u}j_1}, \ldots, \tilde{X}_{\mathbf{u}j_{n_u}}$. Let it be $\tilde{q}_\mathbf{u}$. The DCW is the direction $\mathbf{e_u}$ is defined as $\tilde{Q}_{proj}(\mathbf{u}) = \tilde{q}_\mathbf{u}\mathbf{e_u}$.

In order to state the theoretical properties of the DCW, first define $\tilde{X}_{\mathbf{u}i} = w_{\mathbf{u}}w_{2j_k}X_{\mathbf{u}i}$, for $i = 1, \ldots, n$. Also, consider the following two functions

$$\Psi_{\mathbf{u}}(X, q) = \mathbb{I}_{\{\|X_{\mathbf{u}\perp i}\| \leq \epsilon\}} \left[|\tilde{X}_{\mathbf{u}i} - q| + \|\mathbf{u}\|(\tilde{X}_{\mathbf{u}i} - q) \right],$$

$$g_{\mathbf{u}}(X, q) = \mathbb{I}_{\{\|X_{\mathbf{u}\perp i}\| \leq \epsilon\}} \left[\left(2\mathbb{I}_{\{\tilde{X}_{\mathbf{u}i} \leq q\}} - 1 \right) - \|\mathbf{u}\| \right].$$

Our results are based on the *population level* properties of the functions $\Psi_{\mathbf{u}}(X, q)$ and $g_{\mathbf{u}}(X, q)$, that is, their behavior when we take an expectation of these functions with respect to the measure extended by X. Such properties are not assumed for sample level functions. An extremely easy example where population and sample values differ may be seen in the context of a Binomial (n, θ) random variable Z. Note that the expectation of Z/n is θ, which is a smooth function on $(0, 1)$. However, the sample expectation, i.e., the same functional computed under the empirical distribution function, is just Z/n, which is supported only on discretely many values, and is not a smooth function.

We assume that $\mathbb{E}\Psi_{\mathbf{u}}(X, q)$ is finite for all potential choices of q, and has a unique minimizer, which we call $q_{\mathbf{u}}^*$. This merely states that there is a unique population parameter to estimate. The sample version does not require uniqueness, but that may be enforced, as is traditionally done, by defining the minimizer to be the infimum over all possible values at which the minimum is reached. In this framework, we have the following results:

Theorem 2. *The sample DCW is a consistent estimator of the population DCW, that is $q_{\mathbf{u}} \to q_{\mathbf{u}}^*$ almost surely as sample size $n \to \infty$.*

Theorem 3. *Under the additional* **population** *level conditions that $\mathbb{E}g_{\mathbf{u}}^2(X, q_{\mathbf{u}}^*) < \infty$, and that the function $\mathbb{E}\Psi_{\mathbf{u}}(X, q)$ is twice continuously differentiable at $q_{\mathbf{u}}^*$ with the second derivative H being positive definite, then as $n \to \infty$*

$$n^{1/2}(q_{\mathbf{u}} - q_{\mathbf{u}}^*) = -n^{-1/2}H^{-1}S_n + o_P(1),$$

where $S_n = \sum_{i=1}^{n} g_{\mathbf{u}}(X_i, q_{\mathbf{u}}^*)$. This implies, in particular, that $n^{1/2}(q_{\mathbf{u}}^* - q_{\mathbf{u}}^*)$ is asymptotically Normal, with asymptotic variance $H^{-1}VH^{-1}$ where we have $V = \mathrm{Var}\, g_{\mathbf{u}}(X, q_{\mathbf{u}}^*)$.

The proofs of these results, and other theorems that follow, require considerable mathematical details. We present a very brief sketch of the main line of argument in the supplementary material, and the details can be made available as needed.

4   Data Depth Using the DCW

We consider the support of the feature vector, $\mathcal{X}$ to be a convex set in $\mathbb{R}^p$. For any given point $\mathbf{p} \in \mathcal{X} \setminus \{\mathbf{0}\}$ the support of a feature of an absolutely continuous random vector X with cumulative distribution function F, we define the *data-depth* as $D(\mathbf{p}, F) = \exp(-\alpha_{\mathbf{p}})$, where $\mathbf{u} = \alpha_{\mathbf{p}}\mathbf{p}/\|\mathbf{p}\|$, and $\tilde{Q}_{proj}(\mathbf{u}) = \mathbf{p}$. We extend this to the point $\mathbf{p} = \mathbf{0}$ by defining $D(\mathbf{0}, F) = 1$. This essentially means that $\alpha_{\mathbf{p}} \in (0, 1]$ is the norm of $\mathbf{u}$, which has the same direction as $\mathbf{p}$, such that $\mathbf{u}^{th}$ DCW is exactly $\mathbf{p}$.

The following properties are available for the data depth function $D(\mathbf{p}, F)$:

Theorem 4. *1. $D(\mathbf{p}, F) = 1$ if and only if $\mathbf{p} = \mathbf{0}$.*
2. $D(t\mathbf{p}, F) \leq D(\mathbf{p}, F)$ for all $\mathbf{p} \in \mathcal{X}$, and all $t \in [0, 1]$.
3. The directional derivative of $D(\mathbf{p}, F)$ with respect to $\mathbf{p}$ exists.
4. The depth function $D(\mathbf{p}, F)$ is smooth in te second argument, in the sense that the Gateux derivative exists.
5. $D(\mathbf{p}, F) \to 0$ as $\|\mathbf{p}\| \to \infty$.

Note that the first two properties are the essential properties of a data-depth, while the rest of the results are technical properties that help understand the depth function better.

5   Geometric Classification Technique

One immediate application of the DCW and the related data-depth algorithm is in supervised learning. Consider a feature vector $X \in \mathcal{X} \subseteq \mathbb{R}^p$ for some (potentially high) dimension p, associated with a label $Y \in \mathcal{Y} = \{0, 1, \ldots, K-1\}$. Assumed that the observed data is a random sample $\{(X_i, Y_i) \in \mathcal{X} \times \mathcal{Y} \subseteq \mathbb{R}^p \times \{0, 1, \ldots, K-1\}\}$ of such (X, Y) pairs. Here K, the total number of labels, is assumed known. Without loss of generality, we assume that observations indexed by $S_k = \{i_{k1}, \ldots, i_{kn_k}\}$ share the common label $Y_{i_{kj}} = k$, for any $k \in \mathcal{Y} = \{0, \ldots, K-1\}$. We assume that $\cup_{k=0}^{K-1} S_k = \{1, \ldots, n\}$, and $S_k \cap S_{\tilde{k}} = \varnothing$ whenever $k \neq \tilde{k}$. We denote the n_k feature observations corresponding to S_k by $\mathbf{X}_k = (X_{i_{k1}}, \ldots, X_{i_{kn_k}})$.

We elicit the label of any new or unlabeled feature vector $x \in \mathcal{X}$ by computing its depths with respect to the K different data clouds of observations $\mathbf{X}_k = (X_{i_{k1}}, \ldots, X_{i_{kn_k}})$. We may then choose the label that corresponds to the highest depth value, or construct labels using more complex usage of the K depth values obtained for x. For example, one alternative to simply choosing the highest-depth label would be to not classify an unlabeled observation if the maximum depth is below a threshold, thus paving the way for potentially extending this algorithm to unsupervised and semi-supervised classification problems, which we do not pursue here.

Notice that the geometric learning-based separating boundary between two labeled classes in the feature space is essentially the locus of the points where the data-depths are equal from both the classes. Thus, the separating curves between labeled classes are essentially *isodepth curves*. This notion can be extended to multiple labeled classes easily, and is of independent interest.

Nevertheless, note that in order to classify a new or unlabeled observation, it is not necessary to obtain the entire isodepth curves. The only computation that needs to be performed is to obtain the data-depth *in the direction of the unlabeled observation* relative to each labeled class, which is trivially a parallelizable procedure requiring at most n one-dimensional projections and few other simple computations. This leads to the geometric learning algorithm being very much amenable to active learning as well as online learning of class labels.

It may be noted that existing techniques for supervised learning either impose shape restrictions explicitly or tacitly, or suffer from curse of dimensionality, or are slow, sequential procedures requiring multiple passes through the data. Many methods of supervised learning suffer from multiple of these issues. For example, methods like logistic regression for two or more class labels, as well as linear or quadratic discriminant analysis explicitly make parametric statistical assumptions, which imply strong restrictions on the shape of the allowed data layout. The classical version of these algorithms are inapplicable for data in high dimensions, and require additional assumptions, typically of sparsity of some functional of the feature vector distributions, for usability in big data. Some of these additional assumptions are unverifiable in data. Nearest neighbor rules implicitly make similar assumptions with a choice of metric and tuning parameters, while support vector machines make an implicit choice of allowable geometry using the kernel function. Methods like (multivariate) density estimation and subsequent learning procedures suffer from the curse of dimensionality. Decision trees and ensemble-based methods like classification and regression trees, bagging, random decision forests, boosting require iterative and often sequential computation, and may impose shape restrictions on the data and may suffer from curse of dimensionality depending on the algorithmic details.

6   Data Geometric Feature Selection

The key benefit of the weighted projection quantile (WPQ) and *geometric learning algorithm* (GLA)-based classification is in high dimensional data. However, the application of the algorithm also provides us high accuracy of classification when applied in conjunction with a suitable feature selection algorithm. To retain the speed benefits of the GLA algorithm, it is important that the feature selection can also be done using a quick and fast method. We have used a very simple and quick algorithm for feature selection which we describe in details below.

A very simple feature selection can be done using how well individual features are correlated with the classes. Using a Spearman's correlation can provide us with a very basic and quick feature ranking. However, the problem with this method is that non-monotonic relations of features with the response.

Another method could be constructed based on a measure of distance of distributions of a feature in each class. If a feature is informative, then the distribution of the feature between the classes will have some separation or distance. On the other hand, if a feature is not informative at all, then the distributions would not be substantially separated. We can use the Kolmogorov-Smirnov test statistic $D_x = sup\|F_1(x) - F_0(x)\|$ for feature x for a two class $(1, 0)$ response. The Spearman's correlation and the Kolmogorov-Smirnov test statistic would be significantly correlated. However, Kolmogorov-Smirnov statistic D would be more sensitive to non-monotonic relations than the Spearman's correlation. This can be illustrated by Fig. 3 comprising of a the graph of a highly informative feature and a relatively non-informative feature from the Dexter dataset from the UCI repository, discussed in details later.

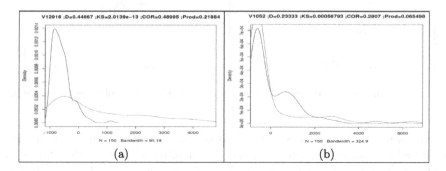

Fig. 3. Plot of two features from the Dexter dataset. The left panel (a) represents a highly informative feature, while the right panel (b) represents a low information feature. Our feature selection protocol quantifies this.

The third scheme of feature ranking can be a constructed using a combination of the two. We used a product of the Kolmogorov-Smirnov distance $\in (0,1)$ with the Spearman's correlation coefficient $\in (0,1)$. We ranked the features in decreasing order based on the product of the Spearman's correlation coefficient and the Kolmogorov-Smirnov test statistic D. Usually a combination of the top few features provide the best predictive ability. So at the second stage we chose a p fraction of features. If there are k features we chose pk features and the response, and constructed a precision matrix, or inverse covariance matrix, of these $pk + 1$ variables together using generalized lasso. Based on the strength of relationship with the response, we re-ranked the features based on the generalized lasso.

Based on the final feature ranking we started with first feature and computed the predictive accuracy on the validation set. Then we progressively included features and kept the feature in the final feature set or deselected a feature based on the increased in prediction accuracy on the validation set. If inclusion of a feature increased the predictive accuracy on the validation set then we kept that feature in the list of selected feature or else we deselected the feature. We continued to do this until the increase in prediction accuracy by inclusion of further features did not benefit the classification.

7   Geometric Learning Example: Simulated Data

In this section we use two simulated datasets to demonstrate the key characteristics of the proposed geometry learning algorithm, and compare its performance with state-of-the-art alternatives.

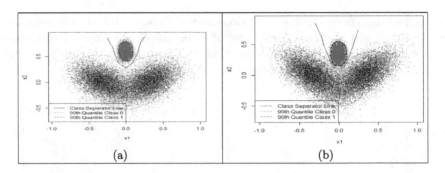

Fig. 4. Isodepth separation curves for a binary classification problem in two datasets

The simulated data contains two classes, and is depicted in Figure 4 with black and red points. We present the two simulated datasets in this figure, in the two panels. In both panels, the black points are clustered in the apple shaped figure, while the red points are in the top leaf cluster. The dotted lines in green and black are the DCW curves corresponding to the black and red observations respectively, corresponding to $\|\mathbf{u}\| = 0.9$. The solid blue curve indicates the iso-depth curve for this dataset, which acts as the curve of separation for the two groups of data. In order to evaluate the classification accuracy and speed of our algorithm, we randomly selected 80 % of the observations from each group for training, and evaluated the geometric learning algorithm on the rest 20 % test data from each group. For comparison purposes, we also used several standard classification algorithms on identical training and testing datasets. This process was repeated 1000 times to ensure sufficient randomization. All results from this section and the next section are based on 1000 randomization runs as described above.

As can be seen from Table 1, the geometric learning algorithm is comparable or better in terms of classification accuracy and speed compared to the state-of-the-art methods. In fact, it is considerably fast in comparison to the other algorithms which achieve similar levels of accuracy in different datasets, of which we have chosen two illustrative examples. In particular, the geometric learning technique is considerably faster than random forest and support vector machines (SVM). In general, parametric methods like linear or quadratic discriminant analysis (LDA, QDA), and logistic regression, or procedures like neural networks do not achieve a good class separator, especially in dataset (b) where the classes are not as distinct as in case of dataset (a).

We used the standard packages and routines in the open source software R, namely glm, lda (MASS), qda (MASS), randomForest, svm (e1071), nnet, knn (class) for the existing supervised learning methods reported here. In the random forest procedure, the number of trees for each run was kept at 500, sampling was done with replacement, and the rest of the parameters were run with default setting. Radial basis kernel $(= \exp(-\gamma|u - v|^2))$ was used for the SVM fit of type *C-Classification* using a scaling of 1 and a class weight equal to the proportion of observation in each class in the train set. For neural nets, the size of the hidden layer was set at 2, case-wise sample weights were set at 1, and entropy fit was used. Note that the random forest algorithm has an inbuilt feature selection capability. Hence, without feature selection random forest tends to perform much better than other comparable algorithms. More details on the classification techniques used for our analysis, and detailed graphical results from this simulation study may found in the supplementary material. Algorithmic and statistical details of these learning techniques can be found in [8]. The standard error of classification accuracy results under any method may be estimated using the relation $s.e. = \sqrt{a(1-a)/1000}$ where is the proportion of accurate classifications, since the results are based on 1000 randomization runs.

8   Performance of the Algorithm on Standard Datasets

We show the performance of the dataset on multiclass multivariate datasets, namely the celebrated Fisher's iris dataset, the colon dataset from the R package cepp, Arcene and Dexter from the UCI Machine Learning dataset library [2] available at https://archive.ics.uci.edu/ml/datasets.html.

8.1   Fisher's Iris Dataset

This is perhaps the best known dataset to be found in the pattern recognition, learning and statistical classification literature. Performance on this dataset is a litmus test for any new proposed machine learning method. The data contains three classes of Iris plants of 50 instances each. The three classes are *Iris setosa*, *Iris versicolor* and *Iris virginica*. There are four features related to the Iris plant and flower in this dataset: sepal length, sepal width, petal length and petal width, all in centimeters. We use the same methods and techniques as in the simulated data analysis discussed above.

In Table 2 we present the average classification accuracy in the test data sets from the Iris data. The running time of all the algorithms are very similar since the dataset is small with 150 observations and 4 predictors. In terms of performance, while all methods perform well, the proposed geometric learning algorithm is the best.

8.2   The Colon Cancer Dataset

The colon dataset is a publicly available dataset for $n = 62$ individuals related to colon cancer. This dataset was generated using Affymetrix oligonucleotide arrays, and contains expressions levels for 40 tumor and 22 normal colon tissues. Out of the originally measured 6500 human genes, $p = 2000$ with the highest minimal intensity across the tissues are selected for classification purposes. Each score represents a gene intensity derived in a process described in [1]. Note that in this case dimension p is considerably higher than sample size n, and thus represents a high dimensional supervised learning problem (Table 3).

It can be seen that the geometric learning algorithm is the most successful among the supervised learning methodologies that were usable for this dataset, both in terms of prediction accuracy, as well as in terms of speed. Methods not reported here were not applicable without additional unverifiable technical assumptions. Support vector machines perform marginally better in terms of accuracy, but requires an 80-fold increase in computing time.

8.3 Arcene Dataset

The Arcene dataset consists of mass spectrometry data obtained with the SELDI technique, and is accessed from the UCI repository [2]. The data contains data from both cancer patients (ovarian or prostate) or healthy individuals. The classification task is to identify the cancer tissues. The data consists of 900 observations and 10000 attributes. Out of these 10000 attributes 7000 real attributes and 3000 artificial random probes.

We use this data to illustrate two things: the speed of the proposed geometric learning algorithm, as well as the efficacy of the proposed fast feature selection procedure. First, we present in Table 4 the results for the geometric learning and several other algorithms, when no feature selection is performed a priori. Notice that the proposed GLA takes negligible amount of time compared to random forest or support vector machine (SVM)-based methodology. It achieves a 82 % accuracy, which is marginally lower than SVM at 84 %, and somewhat lower than 89 % achieved by random forest method which is about 4550-times slower.

We next performed the fast feature selection as discussed in Sect. 6. With the feature selection we were able to improve the accuracy of classification to 0.92 with a balanced error rate (BER) of 0.08603 with 9 features out of 10000. Balanced error rate (BER) is defined as the mean prediction error in all classes. This is better performance the accuracy achieved for the NIPS feature selection competition from which this data originated [5,6]. The mean BER obtained at the NIPS challenge is 0.119 ± 0.012 and the best BER obtained was 0.10. We ran some of the standard classification algorithms, and the results are presented in Table 5. Once a small number of features are selected, classical methods like naive Bayes (discriminant analysis) and generalized linear model (GLM), of which logistic regression is a special case, become viable, and are reported here. This table shows that after feature selection, the proposed method is about three times faster than the sate-of-the-art random forest method, and achieves greater accuracy. None of the classical methods perform as well.

8.4 Dexter Dataset

The Dexter dataset is a text classification dataset with a bag-of-word representation, and is also available at the UCI repository [2]. This is a two class classification problem based on corporate acquisition text collected from Reuters news items. The dataset consists of 20000 attributes. Out of these 20000 attributes, real attributes are 9947 and the rest of the 10053 attributes are random probes for the NIPS feature selection and classification contest. The train set consists

Table 1. Comparison of several supervised learning algorithms on the simulated example, in randomly selected test sets. The classification accuracy is the average proportion of test sets observations correctly classified.

Method	Data (a)		Data (b)	
	Accuracy	Run time	Accuracy	Run time
Geometric learning	0.996	1.46	0.976	1.48
Logistic regression	0.981	0.28	0.901	0.31
LDA	0.969	0.22	0.881	0.20
DA	0.992	0.27	0.974	0.21
Random forest (500 trees)	0.998	23.61	0.976	31.77
Neural network	0.967	4.76	0.962	3.56
SVM	0.984	6.45	0.974	8.54

Table 2. Performance of different classification algorithms on the Fisher iris dataset

Method	Accuracy
Geometric learning	0.9733
LDA	0.9600
QDA	0.9667
Random forest (500 trees)	0.9667
Neural network	0.9267
SVM	0.9533

Table 3. Classification results in the colon cancer data

Method	# Mis-classified	Accuracy	Run time
Geometric learning	9	0.854	0.21
Random forest (500 trees)	10	0.839	226.92
LDA	15	0.758	45.12
SVM	8	0.871	17.36

Table 4. Arcene classification without feature selection

Method	CPU time	Accuracy
Geometric learning	3.67	0.825
Random forest	16714.20	0.895
SVM	966.86	0.842

Table 5. Arcene classification output with feature selection

Method	CPU time	Accuracy
Geometric learning	0.32	0.92
Random forest	0.98	0.90
SVM	0.33	0.91
Naive bayes	0.28	0.89
GLM	0.35	0.74

Table 6. Dexter classification output

Method	CPU time	Accuracy
Geometric learning	0.43	0.89
Random forest	1.23	0.91
SVM	0.51	0.90
Naive bayes	0.37	0.86
GLM	0.35	0.69

of 300 observations and the validation set consists of another 300 observations. With feature selection we achieved a best classification accuracy of 0.89 with a BER of 0.116 which is also comparable to the NIPS performance. The best accuracy was achieved using top 75 features. Results from this analysis is presented in Table 6. The results show that the proposed geometric learning algorithm is again one of the fastest methods, which achieves nearly comparable accuracy with the best existing techniques.

9   Conclusions and Future Directions

The geometric learning algorithm can be seen to be very versatile, and applicable in complicated supervised learning problems, as well as in high dimensions. One future research to pursue is on theoretical quantification of its classification error bound. Another important consideration in this line of research is to evaluate its performance in unsupervised learning problems.

The illustrative examples of the previous section demonstrates several things. First, the proposed procedure captures the geometry of the data reasonably well. Second, owing to its geometric properties, the speed of the algorithm is not particularly affected by the dimensionality of the feature space. However, there is plenty of scope of parallelization, both in dimensions as well as in sample size. Third, our method is robust against lack of assumptions, and seems to be fairly functional even without feature selection in many examples. Naturally, feature selection improves performance, but the rough and fast feature selection procedure suggested here seems adequate, showing additional robustness. In all examples, the geometric learning algorithm is either the best in terms of accuracy, or reasonably close to the best, this showing the speed and robustness does not drastically compromise its efficiency.

Some of the limitations of the geometric learning method arise from the topological restrictions it imposes on the data corresponding to any labeled class. While such restrictions are minimal, it is nevertheless unlikely to succeed in classification problems where the data cloud for any class consists of more than one connected component, or have genus greater than zero. We can envisage how to extend the geometric learning algorithm to address these kind of data features, but nevertheless additional research needs to be carried out.

Acknowledgements. This research is partially supported by NSF grant # IIS-1029711, NASA grant #-1502546) the Institute on the Environment (IonE), and College of Liberal Arts (CLA) at the University of Minnesota.

References

1. Alon, A., et al.: Broad patterns of gene expression revealed by clustering analysis of tumor and normal colon tissues probed by oligonucleotide arrays. Proc. Natl. Acad. Sci. USA **96**, 6745–6750 (1999)
2. Bache, K., Lichman, M.: UCI machine learning repository (2013)
3. Chaudhuri, P.: On a geometric notion of quantiles for multivariate data. J. Am. Stat. Assoc. **91**, 862–872 (1996)
4. Ferguson, T.S.: Mathematical Statistics. A Decision Theoretic Approach. Academic Press, New York (1967)
5. Guyon, I., et al.: Feature selection with the CLOP package. Technical report (2006)
6. Guyon, I., et al.: Competitive baseline methods set new standards for the NIPS 2003 feature selection benchmark. Pattern Recogn. Lett. **28**, 1438–1444 (2007)
7. Haldane, J.B.S.: Note on the median of a multivariate distribution. Biometrika **35**, 414–415 (1948)
8. Hastie, T., Tibshirani, R., Friedman, J.: The Elements of Statistical Learning: Data Mining, Inference, and Prediction. Springer, New York (2009)
9. Mukhopadhyay, N., Chatterjee, S.B.: High dimensional data analysis using multivariate generalized spatial quantiles. J. Mult. Anal. **102–4**, 768–780 (2011)

Constructing Parallel Association Algorithms from Function Blocks

Ivan Kholod[✉], Mikhail Kuprianov, and Andrey Shorov

Saint Petersburg Electrotechnical University "LETI", ul. Prof. Popova 5,
Saint Petersburg, Russia
{iiholod, ashxz}@mail. ru, mikhail. kupriyanov@gmail. com

Abstract. The article describes the method of construction of association rules retrieval algorithms out from function blocks having a unified interface and purely functional properties. The usage of function blocks to build association rules algorithms allows modifying the existing algorithms and building new algorithms with minimum effort. Besides, the function block properties allow to transform the algorithms into parallel form, thus improving their efficiency.

Keywords: Data mining · Parallel data mining · Data mining algorithms · Parallel algorithms

1 Introduction

Data mining algorithms are created at the confluence of various fields of mathematics, computer science: statistics, theory of probability, heuristic algorithm, neural networks, genetic algorithms, etc. Many data mining algorithms emerging from a certain theory, develop, modify and give birth to a whole family of similar algorithms distinguishing themselves through certain blocks. Association rules algorithms of the Apriori family built on a common principle are an example of such algorithms: support of any series of objects cannot exceed minimal support of any of their subsets.

Implementation of such algorithms as solid blocks (not decomposed into separate blocks) hinders their modification and optimization, in particular make the conversion into the parallel form more complicated. An alternative to the solid implementation can be the decomposition of algorithms into separate blocks with further creation of algorithms by way of combining of the said blocks.

The ability to build algorithms of separate blocks is inherent to the functional programming language. They allow to describe the functional expression (algorithm), as a sequence of "pure" function (block) invocations. We offer an approach to building association rules algorithms with the use of functional programming paradigm. This approach not only allows to create new algorithms on the basis of the ones already existing, but also allows to easily convert consecutive algorithms into parallel forms. On the bases of the developed framework association rules algorithms: Apriori and Partition were realized and also their modifications for the concurrent running.

The paper is organized as following. The next section is a review of research in the field of association rules algorithms parallelizing. The Sect. 3 contains the description of algorithm building of function blocks and the description of blocks common to data

P. Perner (Ed.): ICDM 2015, LNAI 9165, pp. 124–138, 2015.
DOI: 10.1007/978-3-319-20910-4_10

mining algorithms. The Sect. 4 describes the Apriori and Partition algorithms developing as a combination of function blocks, including their parallel forms. The fifth section describes implementation of function blocks for the Apriori and Partition algorithms and they parallel variants. This section also deals with experiments with the implemented algorithms with considering of the results.

2   Related Work

Research in the field of parallel and distributed data mining have been also carried out for quite a while. As a matter of fact, separate focus areas can be distinguished within the data mining field [1]: parallel data mining and distributed data mining. There are several problems in developing parallel algorithms for a distributed environment with association discovery data mining which is being considered in research work [2]: data distribution, load balancing, minimizing communication and other. Achieving these goals in one algorithm is nearly impossible, as there are tradeoffs between several of the above points. Existing algorithms for parallel data mining attempt to achieve an optimal balance between these factors.

There are many parallel association rules algorithms in related works. In papers [1] and [2] surveys of parallel association rules algorithms variations are presented. The major paradigms for parallel association rules algorithms are:

Shared Candidates: A common paradigm for parallel association mining is to partition the database in equal sized horizontal blocks, with the candidate item sets replicated on all processors. For Apriori-based parallel methods, in each iteration, each processor computes the frequency of the candidate set in its local database partition. This is followed by a sum-reduction to obtain the global frequency. The infrequent item sets are discarded, while the frequent ones are used to generate the candidates for the next iteration.

This paradigm trades off I/O and duplication for minimal communication and good load-balance: each workstation must scan its database partition multiple times and maintains a full copy of the (poor-locality) data structures used, but only requires a small amount of per-iteration communication and has a good distribution of work. Barring minor differences, the methods that follow this data-parallel approach include PEAR [3], PDM [4], Count Distribution [5], FDM [6], Non-Partitioned Apriori [7], and CCPD [8].

Partitioned Candidates: Here each processor computes the frequency of a disjoint set of candidates. However, to find the global support each processor must scan the entire database, both its local partition, and other processor's partitions (which are exchanged in each iteration).

This paradigm trades off a huge amount of communication (to fetch the database partitions stored on other workstations) for better use of machine resources and to avoid duplicated work. The main advantage of these methods is that they utilize the aggregate system-wide memory by evaluating disjoint candidates, but they are impractical for any realistic large-scale dataset.

Algorithms implementing this approach include three Apriori-based algorithms: Data Distribution [5], Simply-Partitioned Apriori [7], and Intelligent Data Distribution [9].

Partitioned Candidates with Selectively Replicated: A third approach is to evaluate a disjoint candidate set and to selectively replicate the database on each processor. Each processor has all the information to generate and test candidates asynchronously.

This trades off duplication (the same data may need to be replicated on more than one node) and poor load-balancing (after redistributing the data, the workload of each workstation may not be balanced) in order to minimize communication and synchronization. The effects of poor load balancing are mitigated somewhat, since global barriers at the end of each pass are not required. Methods in this paradigm are Candidate Distribution [5], Hash Partitioned Apriori [7], HPA-ELD [7] and PCCD [8], all of which are Apriori-based.

Vertically Partitioned: Here, the data is assumed to be vertically partitioned among processors. After an initial tidlist (list of transaction ID) exchange phase and class scheduling phase, the algorithms proceed asynchronously. In the asynchronous phase each processor has available the classes assigned to it, and the tidlists for all items. Thus each processor can independently generate all frequent itemsets from its classes.

Little synchronization or communication is needed, since each processor can process its partitioned dataset independently. A transformation of the data during partitioning allows the use of simple database intersections (rather than hash trees), maximizing cache locality and memory bandwidth usage. The transformation also drastically cuts down the I/O bandwidth requirements by only necessitating three database scans. This paradigm is used by four algorithms proposed Zaki et al. [10]: ParEclat, ParMaxEclat, ParClique, and ParMaxClique.

The individual approach (when for each association rules algorithm is chosen most efficient parallel structure for the given conditions) the complexity and effort for elaboration of parallel algorithms is very high. At that this effort is aimed at adapting the algorithms to execution strictly in the required conditions. The changes to the conditions lead to the necessity of conversion of the algorithm which is in fact a creation of a new algorithm. Thus a conclusion can be made that the researches carried out at present do not provide a method allowing to modify the data mining algorithms and convert them into the parallel form with minimal effort.

3   Concept of Constructing Data Mining Algorithm as Functional Expression

We offer an approach to constructing parallel association rules algorithms with the use of functional programming paradigm. This approach allows to easily convert sequential algorithms into parallel algorithms for various paradigms. The algorithm can be created from separated blocks if these blocks have the following features: (1) they are interchangeable; (2) they are executed in arbitrary order.

The first feature can be implemented by unifying input and output interfaces of the block. All these blocks must take the same input argument and return the same output result set. The second feature can be implemented similarly as a function in functional programming languages. These languages are based on λ-calculus theory [11]. The theory knows the following Church–Rosser theorem [12]: when applying reduction rules to terms in the lambda calculus, the ordering in which the reductions are chosen does not make a difference to the eventual result. Thus in accordance with the Church-Rosser theorem λ–functions can be executed in any order (and even parallel) because the λ–function has futures of the pure function [12]. We will call the **function block** – the block of algorithm if it has unified interface and is implemented as pure function. A data mining algorithm can be presented as a sequence of function blocks:

$$dma = fb_n {}^\circ\, fb_{n-1} {}^\circ ... {}^\circ\, fb_i {}^\circ ... {}^\circ\, fb_1 = fb_n\,(d,\, fb_{n-1}\,(d,\, ... \, fb_i\,(d,\, ... \, fb_1(d,\, nil)...)..)),$$

where fb_i: is function (function block) of the type $FB:: D \to M \to M$, where

- D: is input data set that is analyzed by the fb_i block;
- M: is mining model that is built by the fb_i block.

Decomposition of any algorithm splits the algorithm into separate logical blocks, cycles, decision, etc. Additionally, data mining algorithms have special blocks: cycle for vectors, cycle for attributes and other. They can also present as functions of the FB type. For example, the conditional operator can be expressed as a function:

$<condition\_function\_name>:: FB$

$<condition\_function\_name>\ m = if\ cf\,(d,\, m)\ then\ fb_t\,(d,\, m)\ [else\ fb_f\,(d,\, m)]^1$, where

- cf: function for calculating the conditional expression;
- fb_t: function of the FB type, which is executed if the result of function cf is true;
- fb_f: function of the FB type, which is executed if the result of function cf is false.

The cf function has type $CF:: D \to M \to Boolean$.

The cycle of an algorithm can be presented using a recursive call of a function:

$<loop\_function\_name>:: FB$

$<loop\_function\_name>\ m = loop'\,(d,\, fbinit\,(d,\, m),\, cf,\, fbpre,\, fbiter)$

$<loop\_function\_name>:: FB$

$<loop\_function\_name>\ m = loop'(\,d,\, fb_{init}\,(d,\, m),\, cf,\, fb_{pre},\, fb_{iter})$

$loop'\ m = if\ cf(d,\, m)\ then\ loop'(d,\, fb_{iter}(d,\, fb_{pre}(d,\, m)),\, cf,\, fb_{pre},\, fb_{iter})$
$$else\ fb_{iter}(d,\, fb_{pre}(d,\, m))$$

- cf: is the function of the CF type determining the condition of a repeated iteration;
- fb_{init}: is the function of the FB type, which initializes the cycle;
- fb_{pre}: is the function of the FB type executed in a cycle prior to execution of the main iteration;
- fb_{iter}: is the function of the FB type of the main iteration function.

Data mining algorithms often have the cycle for vectors and the cycle for attributes. The cycle for vectors can present as function of FB type:

$<for\_all\_vectors\_cycle\_name> = loop'\,(d,\, fb_{initW}(d,\, m),\, fb_{initW},\, fb_{preW},\, fb_{iter})$, where

---

[1] Alternative part (*else*) is optional.

- cf_W: conditional function checks whether all the vectors have been processed;
- fb_{initW}: initialization function performs initialization of a vector counter initializing it by the index of the 1st vector;
- fb_{prevW}: preprocessing function changes the vector counter assigning the index of the next vector.

The cycle for attributes can also present as function of FB type:

$$< for\_all\_attributes\_cycle\_name > = loop'(d, fb_{initA}(d, m), fb_{initA}, fb_{preA}, fb_{iter}), \text{where}$$

- cf_A: conditional function checks whether all the attributes have been processed;
- fb_{initA}: initialization function performs initialization of an attribute counter by initializing this counter with the use of index of the first attribute;
- fb_{prevA}: preprocessing function changes an attribute counter assigning the index of the next vector.

To make the concurrent execution of a data functional expression, it must be converted into the form in which the function blocks will be invoked as arguments. To this end we should add a function which will allow data-parallelizing in the data mining algorithms:

$<parallelization\_function\_name> :: FB$
$<parallelization\_function\_name> = join \circ fb \circ split$ where

- *split:* function fulfilling the splitting of the data set D and of the mining model M and returning cortege of two list: separated parts of data set *[D]* and separated parts of mining model *[M]* (size of list is defined by the number of available computing processor):

  $split :: D \rightarrow M \rightarrow \{[D], [M]\};$

- *join:* function joining the mining models from the list *[M]* and returning the merged mining model *M:*

  $join :: [M] \rightarrow M.$

- *fb:* function block executed concurrently.

  Thus, parallel function can present as:
  $<parallel\_function\_name> = join ([fb_i (split(d, m)[0] [0]), ..., fb_i(split (d, m)[n] [n])]).$

  In the *join* function the elements of the mining model list *[M]* are computed by the fb_i block, the arguments of which are data sets and mining models from the lists builded by the *split* function. Thus according to Church-Rosser theorem [12] reduction (execution) of such functional expression (algorithm) can be doing concurrently.

4 Apriori Algorithms as Functional Expressions

The Apriori algorithm is described in 1994 by Ramakrishnan Srikant and Rakesh Agrawal in paper [13]. Below gives listing iterated view of Apriori algorithm [16].

```
   //create transactions set T from input data set D
1    for all vectors d ∈ D {
2       T={t={d.itemid | d.transactionid = t.tid, t∈T}}
3    }
     //create L₁ = {large 1-itemsets} for transaction set T
4    for all transactions t ∈ T {
5       for all items i ∈ t {
6          i.count ++;
7          L₁ = L₁ ∪{i} | i.count >= minsupport
8       }
9    }
     //create Lₖ = {large k-itemsets}
10   for (k=2; Lₖ₋₁≠∅; k++) {
        //generate new candidates: Cₖ = apriori-gen(Lₖ₋₁)
11      for all itemsets l₁ ∈ Lₖ₋₁ {
12         for all itemsets l₂ ∈ Lₖ₋₁ {
13            if (l₁[0] = l₂[0] & .. & l₁[k-2]=l₂[k-2] &
                                      l₁[k-1] < l₂[k-1]){
14               cₖ = {l₁[0], .., l₁[k-2], l₁[k-1], l₂[k-1]}
                 //Lₖ = { cₖ ∈ Cₖ | cₖ.count >= minsupport}
15               for all transactions t ∈ T {
16                  if(cₖ ∈ t) cₖ.count++
18                  Lₖ = Lₖ ∪{ cₖ  | cₖ.count >= minsupport}
19               }
20            }
21         }
22      }
23   }
     // generate association rules R
24   for (k=2; Lₖ ≠∅; k++) {
25      for all itemsets l ∈ Lₖ {
26         for all items i ∈ l
27            if (l.count/{l/i}.cont >= minconf) R=R∪l/i-> i
30      }
31   }
```

Result of Apriori algorithm is set of large k- item sets. These item sets are used for generation of set association rules. Specification PMML [14] from DMG Group describes result of association rules algorithms as AssociationModel model. It contains lists:

- list of all items: $I::[i]$;
- list of large k-itemset: $L_k :: [l]$;
- list of association rules: $R :: [r]$.

Additional, we added two lists for save middle results of association rules algorithm:

- list of all transactions: $T :: [t]$;
- list of large k-itemsets set $\{ L_1, L_2, ..., L_k \}$: $LL :: [L]$.

Such, mining model AssociationModel can present as cortege: $M :: \{I, T, LL, R\}$.

The Apriori algorithm (rows 4–23) solves only one part of association task: finding of large item sets. We must add yet two phase for fully solving association task:

- the first phase (rows 1–3): building of transaction set T from input data set D (often, input data set D does not contain transactions, but contains records (vectors) as pair $<d.transactionid, d.itemid>$);
- the last phase (rows 24–31): generation of association rules set R from list of all large k-item sets LL.

To present the Apriori algorithm as composition of functional blocks we identified following functions of FB type (function blocks):

- vectors cycle function that that for all vectors calls function:

  - *buildTransaction:* building of transaction $t \in T$ from current vector $v \in D$ (line 2):

  $$vectorsCycle = loop'(d, fb_{initW}(d, m), fb_{initW}, fb_{preW}, buildTransaction)$$

- cycle function that for all items of current transaction ($i \in I$ and $i \in t$) calls a composition of two functions

  - *createLarge1ItemSet:* creating of large 1-itemsets $c_1 = \{i| i \in I, i.count >= minsupport \} \in L_1$ (line 7);
  - *calculate1ItemSetSupport:* calculating support of 1-itemsets $c_1 = \{i| I \in I\} \in C_1$ (line 6);

  $$transactionItemsCycle = loop'(d, fb_{initI}(d, m), fb_{initI}, fb_{preI},$$
  $$createLarge1ItemSet \,^{\circ}calculate1ItemSetSupport)$$

- cycle function that calls function *transactionItemsCycle* for all transaction ($t \in T$) (lines 4–9):

  $$transactionsCycle = loop'(d, fb_{initT}(d, m), fb_{initT}, fb_{preT}, transactionItemsCycle)$$

- cycle function that for all large k-1-itemset from k-1-itemsets ($l \in L_{k-1}$) starting of current large k-1-itemset (out cycle) calls a composition of functions:

  - *createKItemSetCandidate:* creating of k-itemsets $c_k = \{i| i \in I \} \in C_k$ (rows 13, 14);
  - *transactionsCycle:* transaction cycle function that call function:

- *calculateKItemSetSupport:* calculating support of k-itemsets (rows 16, 17);
- *removeUnsupportItemSet:* removing of k-itemsets $c_k \in C_k$ with support less then value of *minsupport* (line 18);

$$k\_1LargeItemSetsFromCurrentCycle = loop'(d, fb_{initT}(d, m), fb_{initT}, fb_{preT},$$
$$createKItemSetCandidate \circ transactionsCycle \circ removeUnsupportItemSet)$$

- cycle function that calls function *k_1LargeItemSetsFromCurrentCycle* for all large k-1-itemset from k-1-itemsets ($l \in L_{k-1}$):

$$k\_1LargeItemSetsCycle = loop'(d, fb_{init}(d, m), fb_{init}, fb_{pre},$$
$$k\_1LargeItemSetsFromCurrentCycle)$$

- cycle function that calls function *k_1LargeItemSetsCycle* for large k-itemsets set starting 2-itemsets to empty k-itemsets ($L_2 ... L_k \in LL$)

$$largeItemSetListsCycle = loop'(d, fb_{init}(d, m), fb_{init}, fb_{pre}, k\_1LargeItemSetsCycle)$$

- cycle function that for all items of current k-itemset ($i \in I$ and $i \in l$) calls function:

- *generateAssosiationRule:* creating association rule $r \in R$ with confidence more value of *minconfidence* r.confidence $>=$ *minconfidence* (line 27):

$$largeItemSetItemsCycle = loop'(d, fb_{init}(d, m), fb_{init}, fb_{pre}, generateAssosiationRule)$$

- cycle function that calls function *largeItemSetItemsCycle* for all large k-itemset from k-itemsets ($l \in L_k$):

$$kLargeItemSetsCycle = loop'(d, fb_{init}(d, m), fb_{init}, fb_{pre}, largeItemSetItemsCycle)$$

- the *largeItemSetListsCycle* function that calls the *kLargeItemSetsCycle* function.

Thus, the Apriori algorithm can present as composition of two the functions:

$$Apriori = largeItemSetListsCycle \circ largeItemSetListsCycle \circ$$
$$transactionsCycle \circ vectorsCycle \qquad (1)$$

In 1995 modification of the Apriori algorithm was suggested – Partition [15]. It splits the data set on parts that have a size enough for an operation memory. For calculating of item set's support in each part of input data it saves with the item set list of transaction IDs (TID) which contains this item set:

$$l_k. <tids > = \{t.tid \mid t \in T, c_k \in t \}.$$

We added two new functions for work with list of TIDs:

- *generationTIDList:* function for creation of list of TIDs with 1-itemset $c_1.\!<\!tids\!>=\{t.tid|\ t \in T,\ c_k \in t\} \in C_1$ and calculating of support as size of the list of TIDs: $c_1.count = |\ c_1.\!<\!tids\!>\!|$;
- *createKItemSetCandidateWithTIDs:* function for creating of k-itemsets with list of TIDs $c_k = \{\{i|\ i \in I\},\{\ t.tid|\ t \in T,\ c_k \in t\ \}\ \} \in C_k$

The *generationTIDList* function must be used instead the *calculate1ItemSetSupport* function in the *transactionItemsCycle* function:

$$transactionItemsCycle = loop'(d, fb_{initI}(d, m), fb_{initI}, fb_{preI},$$
$$createLarge1ItemSet\ °generationTIDList)$$

The *createKItemSetCandidateWithTIDs* function must be used instead composition of functions *createKItemSetCandidate* °*transactionsCycle* °*removeUnsupportItemSet* in the *k_1LargeItemSetsFromCurrentCycle* function:

$$k\_1LargeItemSetsFromCurrentCycle = loop'(d, fb_{initT}(d, m), fb_{initT}, fb_{preT},$$
$$createKItemSetCandidateWithTIDs)$$

Using the parallel function we can transform functional expressions (1) to various parallel forms to create parallel association rules algorithms. For example, we can parallel execute the *vectorsCycle* function. For this, need to add new parallel function:

$$vectorsCycleParall = join\ °vectorsCycle\ °split.$$

Thus, the parallel Apriori algorithm (1) will present as:

$$AprioriVectorsCycleParallel = largeItemSetListsCycle\ °largeItemSetListsCycle°$$
$$transactionsCycle\ °\mathbf{\textit{vectorsCycleParall}}$$

Another example is parallel execution of *largeItemSetListsCycle* function:

$$largeItemSetListsCycleParall = join\ °largeItemSetListsCycle\ °split,$$

then the parallel Apriori algorithm (1) will present as:

$$AprioriLargeItemSetCycleParallel = largeItemSetListsCycle\ °$$
$$\mathbf{\textit{largeItemSetListsCycleParall}}\ °transactionsCycle\ °vectorsCycleParall$$

Thus, we can construct parallel data mining algorithm by parallelization of any function of the functional expression (however not all variations of parallel algorithm will efficient and correct). For example, parallelizing of the *transactionsCycle* block in the Apriori algorithm will give incorrect results, because the support of each item set will be calculated with various transaction subsets (with various parts of general transaction set *T*). To detect incorrect results we need to compare the mining model built by a sequential algorithm with the mining model built by a parallel algorithm.

5   Implementation of Parallel Association Rules Algorithms

We extended Xelopes library [16] (data mining algorithms library) and implemented all functions as classes of an object-oriented language Java. Figure 1 shows the class diagram of these blocks. Here the function of FB type is described by the class *Step*. Since the condition and the cycle are function of FB type, *DecisionStep* and *CyclicStep* classes corresponding to them are inherited from the class *Step*. For the implementation of cycle for vectors and cycle for attributes defined *VectorsCycleStep* class and *AttributesCycleStep* class. To form the target algorithm defined the class *MiningAlgorithm*. It contains a sequence of all steps of the algorithm – steps. In the *initSteps* method occurs formation of algorithm structure by creating of the steps which determining of sequence and nesting of their execution. For parallelization of algorithms implemented *ParallelStep* class. As a step of the algorithm, it also inherits from the Step class. It contains a sequence of steps which shall be executed in parallel and implemented two main split and join methods. Most typical type of parallelization for data mining algorithms is parallelization by data. For parallelization of algorithms by data added *ParallelByData* class. It inherits from the class *ParallelStep*. The *ExecutionHandler* class is adapter for execution of an algorithm's branch. Now the framework contains only one implementation of the handler adapter the *ThreadExecution Handler* class for multithreads parallel execution. The library elements were presented in detail at the conference [17].

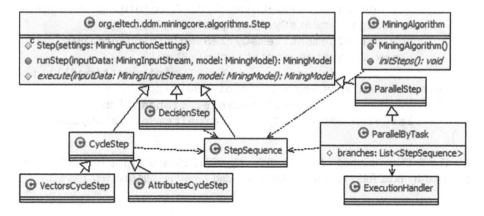

Fig. 1. Class diagram of framework for creating parallel data mining algorithm

To implement association rules algorithms are presented in expressions (1) we implemented new classes inherits from the *Step* class (for function of FB type) and the *CyclicStep* class (for cycle). The *initSteps* method of the *AprioriAlgorithm* class (AA) constructs the Apriori algorithm as functional expression (1) from classes inherited the *Step* class:

```
protected void initSteps() throws MiningException {
    VectorsCycleStep vectorsCycle =
                    new VectorsCycleStep(miningSettings,
                new BuildTransactionStep(miningSettings));
    TransactionsCycleStep transactionsCycle =
                    new TransactionsCycleStep(miningSettings,
        new TransactionItemsCycleStep(miningSettings,
            new Calculate1ItemSetSupportStep(miningSettings),
            new CreateLarge1ItemSetStep(miningSettings)));
    LargeItemSetListsCycleStep largeItemSetListsCycle =
            new LargeItemSetListsCycleStep(miningSettings,
        new K_1LargeItemSetsCycleStep(miningSettings,
        new K_1LargeItemSetsFromCurrentCycleStep(
                                    miningSettings,
            new CreateKItemSetCandidateStep(miningSettings),
            new IsThereCurrenttCandidate(miningSettings,
            new TransactionsCycleStep(miningSettings,
                new CalculateKItemSetSupportStep(
                                    miningSettings)),
            new RemoveUnsupportItemSetStep(
                                    miningSettings)))))); 
    LargeItemSetListsCycleStep largeItemSetListsCycle2 =
            new LargeItemSetListsCycleStep(miningSettings,
        new KLargeItemSetsCycleStep(miningSettings,
        new LargeItemSetItemsCycleStep(miningSettings,
        new GenerateAssosiationRuleStep(
                                    miningSettings))));
    steps = new StepSequence(miningSettings,
                    vectorsCycle, transactionsCycle,
        largeItemSetListsCycle, largeItemSetListsCycle2);
}
```

The *initSteps* method of the *PartitionAlgorithm* class (PA) has two differences (selected by bold font) for *transactionsCycle* and *largeItemSetListsCycle* blocks:

```
TransactionsCycleStep transactionsCycle =
                new TransactionsCycleStep(miningSettings,
        new TransactionItemsCycleStep(miningSettings,
            new GenerationTIDListStep(miningSettings),
            new CreateLarge1ItemSetStep(miningSettings)));
LargeItemSetListsCycleStep lislcs =
    new LargeItemSetListsCycleStep(miningSettings,
        new K_1LargeItemSetsCycleStep(miningSettings,
            new K_1LargeItemSetsFromCurrentCycleStep(
                                    miningSettings,
                new CreateKItemSetCandidateWithTIDsStep(
                                    miningSettings),
                new RemoveUnsupportItemSetStep(
                                    miningSettings)))));
```

To transform a sequenced form of algorithm to parallel form need to wrap parallelized block by the *ParallelByStep* class. For example, for parallel execution the *vectorsCycle* block need to create object of *ParallelByData* class with this block:

```
VectorsCycleStep vectorsCycle =
                new VectorsCycleStep(miningSettings,
        new BuildTransactionStep(miningSettings));
    ParallelByData parallelVectorsCycle =
    new ParallelByData(miningSettings, vectorsCycle);
```

Next need to use the parallelized block. For example, the *parallelVectorsCycle* block:

```
steps = new StepSequence(miningSettings,
            parallelVectorsCycle, transactionsCycle,
        largeItemSetListsCycle, largeItemSetListsCycle2);
```

Thus, we implemented for the Apriori algorithm 3 parallel variations:

- parallel execution of the *vectorCycle* block and splitting of the data set (AVP);
- parallel execution of first *largeItemSetListsCycle* function block and splitting of only the mining model (in part of transactions list) (ALP);
- parallel execution of second *largeItemSetListsCycle* function block and splitting of only the mining model (in part of large item sets list) (ARP).

For the Partition algorithm we also implemented 4 parallel variations:

- parallel execution of the *vectorCycle* block and splitting of the data set (PVP);
- parallel execution of the *transactionsCycle* block and splitting of the data set (PTP);
- parallel execution of first *largeItemSetListsCycle* function block and splitting of only the mining model (in part of transactions list) (PLP);
- parallel execution of second *largeItemSetListsCycle* function block and splitting of only the mining model (in part of large item sets list) (PRP).

We have performed several experiments for the implemented association rules algorithms. The experiments have been performed with various input data sets (Table 1). These data sets contain various numbers of vectors, transactions and items. The average size of a transaction is also different for each data set. The number of association rules and the maximal size of a large item set built from these date sets are also different. Thus, we have used the data sets for which the association rules algorithms work with various loading.

Table 1. Experimental data sets

Input data set	Number of vectors	Number of transaction	Number of items	Avg. size of transaction	Number of rules	Max. size of large item set
T200	1 000	200	10	5	138	3
T2000	10 000	2 000	10	5	90	2
T20000	100 000	20 000	10	5	90	2
I5	5 000	1 000	5	5	90	3
I10	10 000	1 000	10	10	761	4
I15	15 000	1 000	15	15	7309	4
I1020	20 000	1 000	10	20	8148	7
I1030	30 000	1 000	10	30	11253	10
I1050	50 000	1 000	10	50	11253	10

The experiments have been done on a virtual multicore computer the following configuration: CPU Intel Xenon (12 cores), 2.90 GHz, 4 Gb, Windows 7, JDK 1.7. The parallel association rules algorithms have been executed for the numbers of cores equal to 2, 4 and 8, respectively. The experimental results are provided in Table 2.

The experiments show that parallel execution of algorithms for data sets with various parameters is very different. The efficiency of the Apriori algorithm increases with the increase of the number of processors:

- for data sets with a large number of transactions and a small number of rules when the *vectorCycle* block is executed concurrently (AVP);
- for the reverse situation (with a large number of rules and a small number of transactions) when the *largeItemSetListsCycle* block is parallelized (ALP).

The Partition algorithm can be parallelized efficiently only for data sets with a large number of transactions, because at this stage it creates a list of transaction ID for each item. Therefore, the efficiency of the Partition algorithm improves when we parallelize the *vectorCycle* and *transactionsCycle* blocks. At the following stages of this algorithm we do not employ the list of transactions, but we need to know a full list of large item sets. For this reason, parallelizing of the *largeItemSetListsCycle* blocks of the Partition algorithm is not efficient.

Table 2. Experimental results (ms)

Algorithm	T	T200	T2000	T20000	I5	I10	I15	I1020	I1030	I1050
AA	1	91	344	14848	47	484	7914	5168	8426	11297
AVP	2	**52**	**180**	**7284**	31	394	6602	5307	8749	11577
	4	**42**	**139**	**4413**	28	403	7064	5310	8819	11935
	8	**36**	**125**	**1986**	31	381	6745	4973	8356	11238
ALP	2	150	327	15276	**43**	**342**	**1683**	**2867**	**7215**	**8045**
	4	115	258	12487	**39**	**196**	**656**	**1424**	**3970**	**4689**
	8	99	247	10222	**47**	**142**	**312**	**688**	**3236**	**4080**
ARP	2	26	281	16838	56	471	8990	5601	9598	12804
	4	27	286	17443	58	515	8737	5651	9324	12569
	8	32	297	19184	79	582	8129	5153	8255	11303
PA	1	17	328	31182	93	265	2607	2112	5048	12872
PVP	2	**17**	**109**	**11977**	31	94	1947	1772	4439	11564
	4	**17**	**104**	**8378**	31	94	1908	1587	4443	11887
	8	**16**	**95**	**6975**	29	109	2221	1789	4489	12048
PTP	2	**15**	**275**	**22987**	79	220	2846	2115	5241	13447
	4	**10**	**234**	**21782**	62	188	2814	2037	4992	13099
	8	**10**	**187**	**10540**	63	215	2634	1971	5032	13197
PLP	2	17	241	22861	47	201	2712	2054	4785	12609
	4	16	226	20090	62	204	3065	2092	5396	14846
	8	12	218	19807	44	225	3357	2083	4555	15548
PRP	2	110	298	20365	65	201	2753	2095	4956	12636
	4	109	344	19667	47	226	2641	2025	4872	12131
	8	94	381	24561	78	236	2446	1695	4521	8569

6 Conclusion

Thus, presentation of a data mining algorithm as a functional expression makes it possible to divide the algorithm into functions of FB type (function blocks). Such a splitting of data mining algorithms into helps to create new algorithms from existing blocks with less effort or modify the existing algorithms by replacing separate blocks. Additionally, this approach also allows us to easily create parallel algorithms from sequential algorithm by adding special structural elements for parallel execution.

Experiments show that various ways of parallelizing association rules algorithm are efficient for various data sets. Therefore we need to test many possibilities for adaptation and selection of the best parallel structure of association rules algorithm. The approach of constructing parallel association rules algorithms from functional blocks allows us to reduce the costs required for algorithm parallelization. The examples described above show that we have created 3–4 parallel algorithms by changing 9–12 lines of the program code. Having measured experimentally the execution time of various forms of parallel algorithms, we then select the most efficient one and use it for this data set and similar data sets in future.

In future we are going to automate the process of parallel algorithm development and selection of the best algorithm.

Acknowledgments. The work has been performed in Saint Petersburg Electrotechnical University "LETI" within the scope of the contract Board of Education of Russia and science of the Russian Federation under the contract № 02.G25.31.0058 from 12.02.2013. This paper is also supported by the federal project "Organization of scientific research" of the main part of the state plan of the Board of Education of Russia and project part of the state plan of the Board of Education of Russia (task # 2.136.2014/K).

References

1. Zaki, M.J., Ho, C.-T.: Large-Scale Parallel Data Mining, pp. 1–23. Springer, Heidelberg (2000)
2. Paul, S.: Parallel and Distributed Data Mining, New Fundamental Technologies in Data Mining, Funatsu, K. (ed.), pp. 43–54 (2011)
3. Mueller, A.: Fast sequential and parallel algorithms for association rule mining: a comparison. Technical report CS-TR-3515, University of Maryland, College Park (1995)
4. Park, J.S., Chen, M., Yu, P.S.: Efficient parallel data mining for association rules. In: ACM International Conference Information and Knowledge Management (1995)
5. Agrawal, R., Shafer, J.: Parallel mining of association rules. IEEE Trans. Knowl. Data Eng. **8**, 962–969 (1996)
6. Cheung, D., Han, J., Ng, V., Fu, A., Fu, Y.: A fast distributed algorithm for mining association rules. In: 4th International Conference on Parallel and Distributed Information Systems (1996)
7. Shintani, T., Kitsuregawa, M.: Hash based parallel algorithms for mining association rules. In: 4th International Conference on Parallel and Distributed Information Systems (1996)
8. Zaki, M.J., Ogihara, M., Parthasarathy, S., Li, W.: Parallel data mining for association rules on shared–memory multi-processors. In: Supercomputing 1996 (1996)
9. Han, E.H., Karypis, G., Kumar, V.: Scalable parallel data mining for association rules. In: ACM SIGMOD Conference on Management of Data (1997)
10. Zaki, M.J., Parthasarathy, S., Ogihara, M., Li, W.: Parallel algorithms for fast discovery of association rules. Data Min. Knowl. Discov. Int. J. **1**(4), 343–373 (1997)
11. Church, A., Rosser, J.B.: Some properties of conversion. Trans. AMS **39**, 472–482 (1936)
12. Barendregt, H.P.: The Lambda Calculus: Its Syntax and Semantics, of Studies in Logic and the Foundations of Mathematics, vol. 103. North-Holland, Amsterdam (1981)
13. Agrawal, R., Srikant, R.: Fast algorithms for mining association rules. In: Proceedings of the 20th VLDB Conference Santiago, Chile, pp. 487–499 (1994)
14. PMML Specification: Data Mining Group. http://www.dmg.org/PMML-4_0
15. Savasere, A., Omiecinski, E., Navathe, S.: An efficient algorithm for mining association rules in large databases. In: 21st VLDB Conference (1995)
16. Barsegian, A., Kupriyanov, M., Kholod, I., Thess, M.: Analysis of Data and Processes: From Standard to Realtime Data Mining, p. 300. Re Di Roma-Verlag (2014)
17. Kholod, I.: Framework for multi threads execution of data mining algorithms. In: 2015 IEEE NW Russia Young Researchers in Electrical and Electronic Engineering Conference, pp. 74-80. February 2–4 (2015)

Data Mining in Finance

Topic Extraction Analysis for Monetary Policy Minutes of Japan in 2014

Effects of the Consumption Tax Hike in April

Yukari Shirota[1]([✉]), Takako Hashimoto[2], and Tamaki Sakura[3]

[1] Department of Management, Faculty of Economics, Gakushuin University,
Tokyo, Japan
yukari.shirota@gakushuin.ac.jp
[2] Commerce and Economics, Chiba University of Commerce, Chiba, Japan
takako@cuc.ac.jp
[3] Japan Center for Economic Research, Tokyo, Japan
sakura@jcer.or.jp

Abstract. In this paper, we will analyze monetary policy of the Bank of Japan, the central bank of Japan, just after the sales tax hike held in April 2014, through November 2014. The period matches the Japan's turning point to the recession owing to the sales tax hike in April 2014. With the Abe second Cabinet which started in December 2012, the Bank set the "price stability target" at 2 percent. We analyzed the Monetary Policy Meeting minutes by text mining technologies. Especially we conducted a topic extraction using the Latent Dirichlet Allocation model from the Meeting minutes. The extracted topics clearly showed the impact of the sale tax hike. The biggest topic was the economic conditions related topic and its ratio peaks corresponded to the time of Japan's GDP data released. In addition, we found that the money easing policy topic ratio increased after the tax hike and that the economic growth related topic ratio showed the decline after the tax hike.

Keywords: Topic extraction · Dirichlet allocation model · Monetary policy minutes by Bank of Japan · Second Abe cabinet · Consumption tax hike

1 Introduction

When we want to know the governmental monetary policy in Japan, minutes of the Monetary Policy Meeting by the Bank of Japan are the most important documents. The bank of Japan is the central bank of Japan. The Bank's Policy Board decides on the basic stance for the monetary policy at Monetary Policy Meetings. The Policy Board discusses the economic and financial situation and then decides an appropriate guideline for money market operations there. Then, we have been conducting observations of the minutes by topic extraction methods.

Monetary Policy Meetings are held once or twice a month, for one or two days. Monetary policy has a significant influence on the daily lives of the public. Hence, the point of the minutes is just after the meeting published on the web by the Bank of Japan

© Springer International Publishing Switzerland 2015
P. Perner (Ed.): ICDM 2015, LNAI 9165, pp. 141–152, 2015.
DOI: 10.1007/978-3-319-20910-4_11

in both Japanese and English. After 10 years pass of the meeting day, its long version minutes is published on the web, as well as the point minutes. In the paper, we shall analyze the Japanese version of the point minutes, because we think that the original version in Japanese can convey more clearly the discussions contents; the discussions in the Meetings are conducted in Japanese.

In this paper, we will analyze the Monetary Policy Minutes, focusing on the effects of the sales tax hike in April 2014. The second Abe Cabinet started in December, 2012 with the policy titled *Abenomics* [1]. In January 2013, the Bank set the "price stability target" at 2 percent in terms the year-on-year rate of change in the *consumer price index* (CPI) and has made a commitment to achieving this target at the earliest possible time. In addition, at the Meeting held on October 31, 2014, the Bank decided monetary easing by accelerating the pace of increase in the monetary base and by increasing asset purchases and extending the average remaining maturity of Japanese Government Bond (JGB) purchases [2–4]. We analyzed the Monetary Policy Meeting point minutes by text mining technologies. Especially we conducted a topic extraction using the Latent Dirichlet Allocation (LDA) model. The target period is from April 2014 to November 2014. As a result, we found that the sale tax hike from 5 percent to 8 percent had greatly changed the topic ratios of the meeting minutes, compared with topics before the tax hike.

There exist researches analyzing monetary policy-making by using LDA [5, 6]. In [5], Hansen et al. exploit a natural experiment in the Federal Open Market Committee in 1993 particularly using the LDA model to measure the effect of increased transparency on debate. In [6], Moniz and Jong they design an automated system that predicts the impact of central bank communications on investors' interest rate expectations. Their corpus is the Bank of England's 'Monetary Policy Committee Minutes'.

Our research aim is to support readers of the minutes, so that many people including foreign people can instantly understand the decision-making processes and decisions. We also would like to know the policy-makers' interests and views on them. The terms and expressions used in the minutes are difficult to understand, compared with the other Japanese reports. A double negation or triple negation expressions are often used; especially most often used in an uncertain economic status. Rather mild expressions to avoid the radical ones would be used there even in hard economic conditions. Owing to the ambiguous features of Japanese sentences and its vague phrases, it is difficult to extract topics from the minutes of Monetary Policy Meetings. We have continuously conducted the topic extraction from the minutes and published papers in Japanese and just one in English [7]. To handle the difficult expressions and grasp the points, we use various techniques such as bi-gram segmentation on the Japanese expressions. Researches concerning stock price prediction and so forth such as [8, 9] are conducted also in Japan. As far as we are concerned, however, there may be no other research on topic extraction on the minutes in Japanese with the aim of topic extraction.

Through the topic extraction in the paper, firstly, we found that the biggest topic is one concerning economic conditions through the period. The topic movement has two peaks of which times correspond to the GDP bad news were released; they were August and November in 2014. In addition, during the period, there was no major overseas topic; every topic was mainly concerned domestic issues. Among the topics, there is one topic with declining topic ratio. We interpret the topic as the economic growth in Japan. Before the tax hike, we found the topic of "an increase of the growth." The change of topics must have showed the changes of Japanese economic climate. From the analysis, a topic of the BOJ monetary easing policy was also found and the ratio gradually increased, compared with the ratio in April. In Sect. 2, we shall explain the Cabinet's monetary policy titled Abenomics and the economics climates at the time. In Sect. 3, we explain the topic extraction method using the LDA model. Then, in Sect. 4, topic extraction results by the LDA model will be explained. Finally we conclude the paper in the last section.

2 Abenomics

Abenomics refers to the economic policies advocated by Shinzō Abe since the December 2012 general election, which elected Abe to his second term as Prime Minister of Japan. Abenomics is based upon "three arrows" of monetary easing, fiscal stimulus, and structural reforms [1]. Because we are interested in the monetary policy after the tax hike, we shall focus on the first arrow namely, the monetary easing.

In January 2013, the Bank of Japan set the "price stability target" at 2 percent in terms the year-on-year rate of change in the consumer price index (CPI) and has made a commitment to achieving this target at the earliest possible time [2, 3]. Previously, the "price stability goal in the medium to long term" was in a positive range of 2 percent or lower in terms of the year-on-year rate of change in the CPI and the Bank set a goal at 1 percent for the time being. The replacing a "goal" with a "target" and setting that target at 2 percent from 1 percent was conducted in January 2013 [3, 10]. In April 2013, the Board members were drastically replaced under the new Governor of the Bank of Japan. Then the Bank introduced quantitative and qualitative monetary easing (QQE) in April 2013 to achieve the price stability target of 2 percent, with a time horizon of about two years. The Bank continues to purchase a new phase of monetary easing both in terms of quantity and quality. It will double the monetary base and the amounts outstanding of JGBs (*Japanese government bonds*) as well as *exchange-traded funds* (ETFs) in two years, and more than double the average remaining maturity of JGB purchase [10]. We think that the explanation about the monetary policy details is needed to understand the extracted topics. Therefore we shall cite the four policies from [2]:

(a) The Adoption of the "Monetary Base Control"
The monetary base is the "Currency Supplied by the Bank of Japan" and is defined as follows [4]:

Monetary base =

Banknotes in Circulation + Coins in Circulation + Current Account Deposits in the

Bank of Japan

With a view to pursuing quantitative monetary easing, the main operating target for money market operations was changed from the uncollateralized overnight call rate to the monetary base. Specifically, the guideline for money market operations is set as follows: The Bank of Japan will conduct money market operations so that the monetary base will increase at an annual pace of about 60–70 trillion yen.

(b) An Increase in JGB Purchases and Their Maturity Extension

With a view to encouraging a further decline in interest rates across the yield curve, the Bank will purchase JGBs so that their amount outstanding will increase at an annual pace of about 50 trillion yen. In addition, JGBs with all maturities including 40-year bonds will be made eligible for purchase, and the average remaining maturity of the Bank's JGB purchases will be extended from slightly less than three years to about seven years – equivalent to the average maturity of the amount outstanding of JGBs issued.

(c) An Increase in ETF and J-REIT Purchases

With a view to lowering risk premium of asset prices, the Bank will purchase ETFs and Japan real estate investment trusts (J-REITs) so that their amounts outstanding will increase at an annual pace of 1 trillion yen and 30 billion yen respectively.

(d) The Continuation of the QQE (Easing)

The Bank will continue with the QQE, aiming to achieve the price stability target of 2 percent, as long as it is necessary for maintaining that target in a stable manner.

In 2013, *Abenomics* had immediate effects on currency exchange rates and stock prices in Japan. The Japanese YEN (JPY) became about 25 percent lower against the U.S. dollar in the second quarter of 2013 compared to the same period in 2012, with a highly loose monetary policy being followed, and by May 2013, the stock market had risen by 55 percent. Then in 2013, US dollars appreciated in the foreign exchange markets, rising 21 percent against JPY. However, owing to the impact of 2014 April's sales tax hike, the GDP showed a consecutive decline in August and November. The Cabinet Office released GDP date on 8th December, 2014 of "Development of Real GDP (Quarterly, Annualized), seasonally adjusted series" are −6.7 percent for April through June and −1.9 percent for July through September[1]. The consecutive decline of GDP were greatly surprising us. Then we shall analyse the effects of the sales tax hike by the topic extraction.

3   Latent Dirichlet Allocation Model

In the section, we shall explain the LDA model we used for the topic extraction. The LDA is a widely-used multi-topic document model based on Baysian inteference method [11]. We shall explain the frame simply. In the LDA model, each topic is supposed to have a set of related words. One document is supposed to have several topics. The topic distribution for document i may be (0.7, 0.3) (in the case of an economic related document) or (0.2, 0.8) (in the case of a disaster related document). To express the possible various distributions, we use Dirichlet distribution by using a

---

[1] cited from http://www.esri.cao.go.jp/jp/sna/data/data_list/sokuhou/files/2014/qe143_2/pdf/gaiyou 1432.pdf.

hyper parameter α. On the same way, we define per-topic word distribution by Dirichlet distribution by using another hyper parameter β. The used symbols are as follows:

α is the parameter of the Dirichlet prior on the per-document topic distributions,
β is the parameter of the Dirichlet prior on the per-topic word distribution,
θ_i is the topic distribution for document i,
ϕ_k is the word distribution for topic k,
z_{ij} is the topic for the jth word in document i, and
w_{ij} is the specific word.

The w_{ij} are the only observable variables, and the other variables are latent variables. ϕ is a Markov matrix of which size is $K \times V$ (V is the dimension of the vocabulary) each row of which denotes the word distribution of a topic. The LDA generative process for a corpus D consisting of M documents each of length N_i is as follows where K denotes the number of topics:

1. Choose $\theta_i \sim$ Dir(α), where $i \in \{1, ..., M\}$ and Dir(α) is the Dirichlet distribution for parameter α
2. Choose $\phi_k \sim$ Dir(β), where $k \in \{1, ..., K\}$
3. For each of the word positions i, j, where $j \in \{1, ..., N_i\}$, and $i \in \{1, ..., M\}$

    (a) Choose a topic $z_{ij} \sim$ Multinominal(θ_i).
    (b) Choose a word $w_{ij} \sim$ Multinominal(ϕ_{zij}).

The multinomnal and Dirichlet distriutions are defined in machine learning textbooks. We want to obtain an estimate of Z that gives high probability to the words that appear in the corpus. z_{ij} represents the topic for the jth word in document i. This problems becomes a maximum a posteriori estimation of $P(W, Z, \Theta, \Phi \mid \alpha, \beta)$. By an integration concerning θ and ϕ, the expression becomes a simple one, P($W, Z \mid \alpha, \beta$). Therefore, we want to obtain Z so that P($Z \mid W, \alpha, \beta$) is maximum. The W is given data. The cost of the calculation is too high because the estimation space size is the number of topics (K) to the power of the dimension of the vocabulary (V), K^V. Namely each word has K options independently. So instead of that, a random walk search method by Gibbs sampling is widely used [12]. We used R [13] and the R packaged offered by "The Comprehensive R Archive Network" abbreviated as CRAN titled "lda: Collapsed Gibbs sampling methods for topic models" developed by Jonathan Chang[2].

4   Topic Extraction Results

In this section, the topic extraction results by LDA model is presented.

We have conducted the monthly topic extraction by the LDA model. Figure 1 shows the topic ratio time series changes. First, the LDA model generates the word frequencies for each topic for the whole eight month text data. Then each month text data is input to the generated total model so we can obtain each month's topic ratio.

---

[2] CRAN web page: http://cran.r-project.org/.

At a time, the total of the ratios is supposed to be one, 100 percent. In LDA model, we have to decide in advance the number of topics. In this work, we decided the number to be seven after some experimentation, because with the <u>seven</u> topics, the economic conditions topic clearly appeared. Table 1 shows the most frequently appeared terms (words) of each topic. From the term distributions and time series changes of the topic ratio, we interpreted the topics in the following ways.

(1) Conditions: economic conditions
(2) Recovery: economic recovery
(3) Prices: commodity prices
(4) Policies: monetary policies by BOJ
(5) Growth: economic growth
(6) Inflation Target: inflation target
(7) Others

Concerning the last topic which has the smallest ratio, we cannot interpret the topic title because the number of terms is so small. We shall explain and interpret the six topics. First, we would like to examine the "Conditions" topic which has the largest ratio among the six topics.

No 1: Conditions: Economic Conditions

We interpreted the implications of the topic as Japan's economic conditions. The terms of the topic are "finance", "consumption", "increase", and "tax rate". <u>The topic ratio change has two tops in August and November</u>, which correspond to the times when the GDP data were released by the Cabinet Office. Then the Cabinet Office reported that Japan's GDP real growth rate compared with the previous quarter shrank an annualized 6.8 percent in the three months from April through June. Next it was released that GDP shrank an annualized 1.6 percent from July to September. That was a second consecutive decline that matched the textbook definition of a recession [14]. Although there are no word "recession" in the Policy Minutes, the Board members must have been concerned about that. We think that the GDP data were lower than they expected. The main reason of the GDP sank is considered to be the April's consumption tax hike.

The two peaks in the topic ratio movement show the released GDP data's impact on the Policy Meetings (See Fig. 1). Before the release in November, in October, the board made a significant decision that they would surprisingly expand a monetary easing program unprecedented in scale. The vote on the decision was five against four; namely the decision was made by a short head. The reason of the decision is because then the <u>inflation forecasts drifted lower</u> and further away from the 2 percent target. When we see the topic movement of ours, as shown in Fig. 1, in October the topics "Inflation target" and "Prices" were surely high.

Back to the topic "Conditions", compared with the October's topic ratio, this topic ratio in November jumped again. We think that the increased ratio showed the greatness of the impact concerning the GDP decline.

No 2: Recovery: Economic Recovery

We interpreted the topic as economic recovery. The word list includes "recovery", "improvement of employment", and so forth. In addition, we found often the following expressions in the minute text *"recover moderately as a trend"* or similar ones.

Table 1. The most frequently appeared word list of topics.

Conditions		Recovery		Prices		Policies		Inflation target		Growth	
finance	565	economics	513	commodity prices	521	mild	350	industory	329	rising	281
consumption	434	demand	284	last minute	184	market	281	JPY	230	growth	215
increase	389	recovery	268	overseas	109	easing	226	stedy	201	hike	175
improve	280	employment	175	capacity (investment)	108	policies	169	target	112	qualitative (easing)	128
investment	225	expectation	77	left amount	90	funds	90	BOJ	98	publication	91
kickback	196	labour	67	decrease	88	risk	86	effect	89	production	58
income	165	monetary base	62	satisfaction	82	flattering	85	continuance	85	pick up	54
tax rate	160	export	60	profit	61	growth	83	asset	78	mind	52
conditions	136	maintain	59	robust	57	finance	64	slow down	65	ASEAN	44
interest rate	116	domestic	50	financial administration	56	maintain	60	good	54	foods	43
export	90	enlarge	49	supply	55	quantitative	50	decrease	49	corporate bond	38
prices	84	target	49	expectation	54	overseas	50	household finance	45	April	38
USA	70	leding	48	June	52	short period	48	China	41	region	37
real	66	governetn	44	production	50	decrease	45	budget	33	strategy	30
March	60	emerging	43	momentum	50	JGB	43	rate	32	Japan	29
demand/supply	55	investigation	37	minus	35	weaker (JPY)	42	goods	29	decision	28
fresh product	43	decline	35	capital	34	goods	42	foreign demand	27	down	28
bank	42	high	34	downstroke press downstroke press	34	cost	32	flexible	35	Europe	27
housing	39	additonal	32	Kuroda (head of BOJ)	30	GDP	30	thinking	26	euro	26
accomplishment	39	execution	28	CP	28	deminish	29	long	24	left	25

For example, in the Minutes we found the following sentences; *With regard to economic activity, members concurred that the economy continued to recover moderately as a trend, although some weakness remained, particularly on the production side, due mainly to the effects of the subsequent decline in demand following the front-loaded*

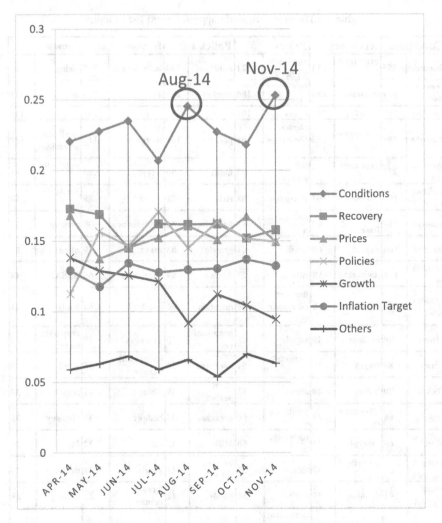

Fig. 1. The topic ratio changes generated by LDA model.

increase prior to the consumption tax hike (cited from the Nov. 18–19 Monetary Policy Point Minutes of BOJ).

The phrase repeatedly appear in the minutes almost constantly through the eight months.

No 3. Prices: Commodity Prices

We interpreted the topic as "commodity prices" mainly because the most frequently word is "commodity prices". The second frequent appeared word is "last-minute" which means the front-loaded increase demand prior to the April's consumption tax hike. The word "tax rate" is also found in the word list. Hence, we think that the impact of the front-loaded increase demand on commodity prices is discussed on this topic.

No. 4. Policies: Monetary Policies by BOJ

The word list of the topic include "easing", "policies", "quantitative", and "maintain". In 2014, one of the most significant economic index change was on the currency exchange rate of JPY against US dollars. The term "weaker (JPY)" appeared only in the topic. We think from the words that this topic concerns the quantitative and qualitative monetary easing (QQE) policy. As mentioned in Sect. 2, the monetary easing policy was introduced in April 2013 and they made the decision in October 2014 that the QQE policy was maintained and expanded.

After May, the topic ratio had increased. We think that this is because the discussions on the policy continuance became more.

No. 5. Growth: Economic Growth

During the eight month period, the topic ratio gradually decrease and in August when the bad GDP data was released the ratio was the lowest. The word list includes "rising", "growth", "hike", and "pick up". Many words of the topic seem to give positive sentiment. Hence, we interpret this topic as the economic growth in Japan.

Before the tax hike, in 2013, the GDP data (Fig. 2) cited from the data of JCER (Japan Center of Economic Research) showed moderate recovery of economic activity and consumption price stability. In April 2014, however, the economic climate has been changed owing to an impact of the tax hike. The Indexes of Business Conditions (CENT'CIDX10C) released from the Cabinet Office, Government of Japan are also

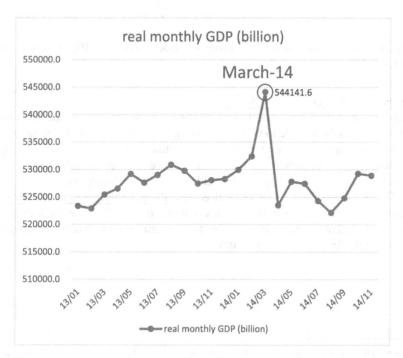

Fig. 2. Time series of seasonally-adjusted real monthly GDP in Japan (from JCER data). The unit of y-axis is billion.

showing the decline of economic conditions (Fig. 3)[3]. The indexes of business conditions are summary measures for aggregate economic activity. They are designed to be a useful tool for analysing current conditions, and for forecasting future economic conditions [15]. As the economic indexes show the slowdown of Japan's economic growth, the topic concerning the Japan's economic growth also decreased.

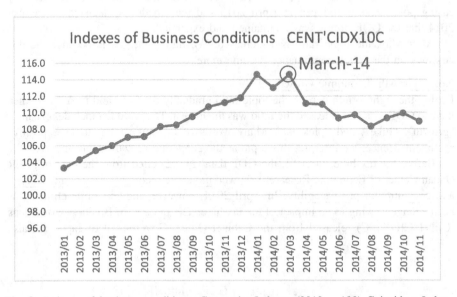

Fig. 3. Indexes of business conditions, Composite Indexes (2010 = 100) Coincident Index, CENT'CIDX10C.

No. 6. Inflation Target: Inflation Target
The movement of the topic ration has no drastic changes and keeps some ratio during the period. We make the title to be "inflation target." This is because the term "target" was frequently appeared and the word list includes "asset", "slowdown", "good (commodity)", and "household finance." The topic ratio keeps almost constant.

As above shown in Sect. 2, the BOJ decided the inflation target to 2 percent in April 2013. The most important policy of BOJ is this "Price Stability Target of 2 Percent" that BOJ set for a regime change of the economic policy [10]. We think that, concerning the inflation target, economic climate on terms such as "business investment" and "household finance" seem to be talked.

---

[3] Cabinet Office, Government of Japan: Composite Indexes (2010 = 100) Coincident Index, CENT'CIDX10C, Indexes of Business Conditions November 2014 Preliminary Release (January 9, 2015).

5   Conclusion

We analyzed the monetary policy minutes of BOJ during the period just after the sales tax hike in April 2014 through November 2014. Our goal in the paper was to analyze the impact of the sales tax hike on the minutes from a neutral viewpoint by a topic extraction method. First, we found the topic concerning the economic conditions which has the biggest topic ratio through the period. The topic movement has two peaks of which times correspond to the GDP bad news were released; they were August and November. The Board meeting members must have discussed the GDP data and recovery measurements of this in the meetings. We can infer that the GDP releases seemed to stimulate the board meetings. In addition, during the period, there was no overseas topic; every topic mainly concerned domestic issues. This result corresponds to the fact there was no events overseas that gave great impact on Japan. Every topic handles domestic issues. Among the topics, there is just one topic which shows the decline movement. We interpret the topic as the economic growth in Japan. Before the sales tax hike, we could see the increase of the growth related topic. However, as the GDP and index of business conditions show, the topic concerning economic growth showed the decline. This decrease corresponds to the slowdown of Japan's economy.

The topic of the BOJ monetary easing policy was found and the ratio increased, compared with that in April 2014. The reason seems to be that the discussion about the continuance of the policy must have been conducted actively.

We think that text mining is very effective when we find an unexpected fact from the results. For example, we found the correlation between bird-flu and declining sales of digital cameras in Japanese market data [16]. Compared with such an unexpected topic finding, the topic extracted from the monetary policy minutes was within our expectation and imagination. It may be only an ascertainment of the newspaper reports or so. However, it is so significant for us to survey the policy-makers' expressions from a neutral standpoint and to know their decisions and the decision-making processes. The translation of the complicated and difficult expressions to plain Japanese expressions would be needed so many people can understand that. In addition, the topic extraction and topic ratio movement would help us feel the changes of economic problems we confront. We will continuously observe and analyze the Policy Minutes by using topic extraction methods.

Acknowledgment. We thank Hirohiko Okumura, an emeritus professor of Gakushuin University for thoughtful comments that improved this paper. We thank Prof Tetsuji Kuboyama for his wide range of knowledge about machine learning that helps our research. This research was partly supported by funds from Gakushuin University Computing Centre as its special research project in 2014.

References

1. Lexicon, F.T.: Definition of Abenomics, 28 January 2014
2. BOJ: The Price Stability Target Under the Framework for the Conduct of Monetary Policy, 22 January 2013 and 10 November 2014. https://www.boj.or.jp/en/announcements/release_2013/k130122b.pdf
3. Cabinet Office, Ministry of Finance, BOJ: Joint Statement of the Government and the Bank of Japan on Overcoming Deflation and Achieving Sustainable Economic Growth, 22 January 2013. https://www.boj.or.jp/en/announcements/release_2013/k130122c.pdf
4. Research and Statistics Department, BOJ: Explanation of "Monetary Base Statistics", May 2013. https://www.boj.or.jp/en/statistics/outline/exp/exbase.htm/
5. Hansen, S., McMahon, M., Prat, A.: Transparency and deliberation within the FOMC: a computational linguistics approach. CEP Discussion Papers, CEPDP1276, Centre for Economic Performance, London School of Economics and Political Science, London, UK (2014)
6. Moniz, A., de Jong, F.: Predicting the impact of central bank communications on financial market investors' interest rate expectations. In: Presutti, V., Blomqvist, E., Troncy, R., Sack, H., Papadakis, I., Tordai, A. (eds.) ESWC Satellite Events 2014. LNCS, vol. 8798, pp. 144–155. Springer, Heidelberg (2014)
7. Shirota, Y., Hashimoto, T., Sakura, T.: Extraction of the financial policy topics by latent dirichlet allocation. In: TENCON 2014-2014 IEEE Region 10 Conference, pp. 1–5 (2014)
8. Schumaker, R.P., Chen, H.: A discrete stock price prediction engine based on financial news. Computer 1, 51–56 (2010)
9. Bollen, J., Mao, H., Zeng, X.: Twitter mood predicts the stock market. J. Comput. Sci. 2(1), 1–8 (2011)
10. BOJ: Price Stability Target of 2 Percent and "Quantitative and Qualitative Monetary Easing", 10 November 2014. https://www.boj.or.jp/en/mopo/outline/qqe.htm/
11. Blei, D.M., Ng, A.Y., Jordan, M.I.: Latent dirichlet allocation. J. Mach. Learn. Res. 3, 993–1022 (2003)
12. Griffiths, T.L., Steyvers, M.: Finding scientific topics. Proc. Natl. Acad. Sci. 101(Suppl. 1), 5228–5235 (2004)
13. R Core Team: R: A Language and Environment for Statistical Computing. R Foundation for Statistical Computing, Vienna, Austria (2013). ISBN 3-900051-07-0, http://www.R-project.org/
14. Bloomberg: Japan's economy takes a dive as Abe weighs delay of second tax hike. In: The Japan Times, 18 November 2014
15. Cabinet Office, GOJ: 14 January 2015. http://www.esri.cao.go.jp/jp/stat/di/di.html
16. Kuboyama, T., Hashimoto, T., Shirota, Y.: Consumer behavior analysis from buzz marketing sites over time series concept graphs. In: König, A., Dengel, A., Hinkelmann, K., Kise, K., Howlett, R.J., Jain, L.C. (eds.) KES 2011, Part II. LNCS, vol. 6882, pp. 73–83. Springer, Heidelberg (2011)

Estimating Risk of Dynamic Trading Strategies from High Frequency Data Flow

Yuri Balasanov[1]([✉]), Alexander Doynikov[2], Victor Lavrent'ev[2], and Leonid Nazarov[2]

[1] University of Chicago, Chicago 60637, USA
ybalasan@uchicago.edu
[2] Lomonosov Moscow State University, Moscow 119991, Russia

Abstract. We consider the problem of risk management in the framework of low latency trading. We suggest an efficient method of real-time analysis of massive data flow from the market. The result of the analysis is a new risk measure Dynamic VaR (DVaR) for risk management of low latency trading robots. The work of DVaR is illustrated on a test example and compared with Traditional VaR and ex-post measure commonly used in high frequency trading.

1 Introduction

Financial markets have always presented researchers with Big Data type problems due to several reasons: data collection from multiple sources in different formats and of different qualities; requirement of the most advanced technology at the time; large number of data points as well as complex relationships between price indices and essentially non-linear and non-Gaussian nature of observations [7,8,13,14].

But the new era of really big data started with the spread of high frequency trading (HFT) when in 1998 SEC authorized electronic trading.

HFT gained significance in early 2000s as a market making approach. In 2005 SEC consolidated National Market System (NMS), authorized by Congress in 1975, into Regulation NMS which was implemented in 2007, a year when HFT started growing really fast.

The real key for HFT is not as much the frequency, not even latency of trading, but algorithmic trading between computers with speed that leaves no chance for human to interfere.

Appearance of trading robots gave rise to completely new types of financial data analyses: personal instincts of a floor trader need to be formalized and implemented utilizing high computing power; terabytes of high frequency price records with millions of events per day need to be analyzed to extract insights for making trading decisions; the problem of market risk management gets a new dimension with significant emphasis on timing and frequency of market events in addition to traditional emphasis on the price impact.

P. Perner (Ed.): ICDM 2015, LNAI 9165, pp. 153–165, 2015.
DOI: 10.1007/978-3-319-20910-4_12

Such significant paradigm shift in financial data analysis requires reevaluation of some core concepts in financial mathematics including the basic model for market price.

In this article we use suitable approach to modeling market price that goes back to [3,4] and further developed in [11] to extend the traditional market risk measure called value at risk (VaR) to account for frequency of events affecting the value of the portfolio.

The obtained results show in particular how the data flow of market prices needs to be organized into the flow of "sufficient statistics", i.e. the frequencies of events of certain types in order to process the risk measurement efficiently in real-time with very low latency.

The rest of the article is organized in the following way. In the introduction we explain the evolution of models for market price that leads to the model that is used in this study and describe the general problem of market risk measurement.

In Sect. 2 we give mathematical summary of the results including the description of the new market risk measure Dynamic VaR (DVaR).

In Sect. 3 we illustrate the work of DVaR and discuss the organization of the market risk measurement process.

1.1   Evolution of Market Price Models

Bachelier Model. The first modern description of the process of price was, probably, given by Bachelier [1]. He used Brownian motion process (later studied by Einstein [5] and rigorously defined by Wiener in early 1920s):

$$S(t) = S(0) + \mu t + \sigma W_t,$$

where W is standard Wiener process i.e. the process with independent increments starting from zero with $\mathbb{P}\{W_t$ is continuous $\} = 1$; $(W_{t+h} - W_t) \sim \mathbb{N}(0, h)$.

The model can be justified by dividing any time interval $[t, T]$ into n equal subintervals $[t_0 = t, t_0 + \tau, t_0 + 2\tau, \ldots, t_0 + n\tau = T]$. Then

$$S(T) - S(t) = \sum_{k=1}^{n} S(t + k\tau) - S(t + (k-1)\tau) = \sum_{k=1}^{n} \Delta S_k. \tag{1}$$

If price increments on subintervals are independent [8] and identically distributed, then according to the central limit theorem (CLT), $S(T) - S(t)$ is asymptotically normal, as $n \to \infty$, as long as $\mathbb{V}[S(t + k\tau) - S(t + (k-1)\tau)] < \infty$ (Lindeberg's condition, 1922).

Non-Gaussian Reality. It became clear soon that observed market price behavior contradicts the Gaussian distribution hypothesis.

In 1960s Fama [7,8] and Mandelbrot [13,14] suggested explanations for non-Gaussian behavior of market prices.

Initially attention was focused on the Lindeberg condition which in more broad sense means that (1) is not dominated by any finite number increments

ΔS_k and in particular, when increments are identically distributed, it means $\mathbb{V}[\Delta S_k] < \infty$.

Mandelbrot noticed extreme variability of second empirical moments of financial data, which could be interpreted as nonexistence of the theoretical second moments, i.e. Lindeberg's condition does not hold. With (1) still in place and i.i.d. ΔS_k, Mandelbrot suggested to model market prices using limit theorems different from CLT, where limit distributions are from the family of stable distributions. Fama showed agreement of the observed market prices with stable distributions.

Stable Distributions. Distribution Function $G(x)$ is Stable if for any a_1, $a_2 \in \mathbb{R}$, $b_1, b_2 > 0$ there exist $a \in \mathbb{R}$ and $b > 0$, such that

$$G(b_1 x + a_1) * G(b_2 x + a_2) = G(bx + a), x \in \mathbb{R}$$

P. Levy proved that distribution function $\mathbb{F}(x)$ can be a limit distribution for the sum $S_n = (X_1 + \ldots + X_n - a_n)/b_n$ of i.i.d. random variables X_i and some $a \in \mathbb{R}$, $b > 0$ iff $\mathbb{F}(x)$ is stable.

Various functional limit theorems establish convergence of discrete time random walks with unlimited variance to stable Levy processes.

Stable processes are fractals: trajectories observed with different frequencies belong to the same stable law, can differ only by the scale parameter.

Unfortunately, there are only 4 known stable distributions with probability densities expressed in terms of elementary mathematical functions (Gaussian, Cauchy, Levy and symmetrical Levy distributions).

Both Bachelier and Mandelbrot approaches still require structure (1) with the possibility of $n \longrightarrow \infty$.

Processes with Stochastic Time. It has been long known that intensity of trading does not remain constant.

Structural assumption (1) requires that on any time interval there is an arbitrary large number n of equidistant moments when price can be observed.

Such assumption might not always work because of random time intervals between the registered trades when the price cannot be measured, or because intensity of trading changes too fast.

Clark [3,4] suggested this as an explanation why CLT does not work with market prices. He described the process of market price as a subordinated Wiener process $W(X(t))$ where W is Wiener process and subordinator $X(t)$ is a stochastic process of "operational time" with non-decreasing trajectories starting from zero. Subordinator turns the Wiener processes into a non-Gaussian one even if increments ΔS_k satisfy CLT.

The condition of CLT that is violated in this case is the determinism of number of terms in the sum (1).

Following this approach, we replace (1) with

$$S\left(T\right) - S\left(t\right) = \sum_{k=N(t)+1}^{N(T)} S\left(k\right) - S\left(k-1\right) = \sum_{k=N(t)+1}^{N(T)} \Delta S_k \qquad (2)$$

where ΔS_k satisfy CLT, the process $N\left(t\right) \geq 0$ starting from zero is the number of trades registered up until t.

If the trading conditions are such that intensity of trading is high enough and/or trading pace is stable during long enough periods of time, then there is a possibility that CLT holds at least locally in time.

Under such conditions eliminating the issue of random summation by looking at blocks of data with large enough sizes seems the most reasonable approach.

This approach in effect leads to using operational time instead of calendar time and naturally results in various stochastic volatility models of price process.

Following this direction there has been a significant theoretical work done that allows approximating the price process with continuous Gaussian martingales including some powerful methods for analyzing implied and realized volatility (see, for example, [15, 16]).

However, in this article we adopt the approach based directly on (2).

Such approach originating in [3, 4] was further developed in [11].

1.2 Risk Measurement

Risk management of a portfolio of financial assets usually is based on one of key risk measures, e.g. VaR or Expected Shortfall, which one can derive from the distribution of loss. Methods of estimation of such distribution can be classified into two major types: (i) application of extreme value theory to simulated sample of portfolio returns [6]; (ii) worst loss analysis for a set of stress scenarios [2].

Both types assume that until the time horizon the portfolio remains unchanged while losses result from adverse changes of market variables. Unfortunately, such assumption is not realistic in low latency case when portfolio immediately adjusts to market changes and the object of risk management is not portfolio itself, but its dynamic response to the market events, i.e. the strategy.

Most importantly, traditional methods do not account for intensity of portfolio changes.

2 Mathematical Summary

In this section using limit theorems for compound Cox processes [11] we derive a real-time risk signal which we call Dynamic VaR (DVaR). In the following section we apply DVaR to the example of a trading system breakdown.

2.1 Definitions

Cox Process. Let $\mathcal{N}$ be a set of all right-continuous non-decreasing integer-valued functions $\nu\left(t\right)$, such that $\nu\left(0\right) = 0$.

Definition 1. *Random process $N(t)$ with trajectories from $\mathcal{N}$ is called point process.*

Let $\mathcal{A}$ be a set of right-continuous non-decreasing functions A_t, such as $A_0 = 0$, $A_t < \infty$ for each $t < \infty$.

Definition 2. *A point process $N_A(t)$ is called a non-homogeneous Poisson process with intensity measure $A_t \in \mathcal{A}$ if $N_A(t)$ has independent increments and $N_A(t) - N_A(s)$ has Poisson distribution with mean $A_t - A_s$.*

Let $\Lambda_t, t \geq 0$, be a random process with trajectories from $\mathcal{A}$.

Cox process is a generalization of non-homogeneous Poisson process in which intensity measure can be stochastic in a certain way.

Definition 3. *A point process $N_\Lambda(t)$ is called Cox process with random intensity measure Λ_t if for any realization A_t of Λ_t the process $N_\Lambda(t)$ is a non-homogeneous Poisson process with intensity measure A_t.*

Definition of Cox process means that we can generate Cox process by first generating a trajectory of intensity measure A_t and then generating trajectory of $N_\Lambda(t)$ as a trajectory of non-homogeneous Poisson process with intensity measure A_t.

If $N_1(t)$ is a homogeneous Poisson process with unit intensity independent of random intensity measure Λ_t then Cox process $N_\Lambda(t)$ is a superposition of $N_1(t)$ and $\Lambda(t)$: $N_\Lambda(t) = N_1(\Lambda(t))$, $t \geq 0$.

Compound Cox Processes. Cox processes allow us to describe randomness of intensity of Poisson process.

With compound Cox processes we can also allow for randomness in the jump sizes of Poisson process or impacts.

Let $X_1, X_2, \ldots$ be i.i.d. random variables with at least two moments $\mathbb{E}[X_j] = a$, $\mathbb{V}[X_j] = \sigma^2, 0 < \sigma^2 < \infty$. Let $N_t = N(\Lambda_t)$ be a Cox process independent of $X_1, X_2, \ldots$ for any $t \geq 0$.

Then the process

$$S_t = \sum_{j=1}^{N(\Lambda_t)} X_j, t \geq 0$$

is called **Compound Cox** process ($\sum_{j=1}^{0} = 0$).

If $\Lambda_t \equiv \lambda t, \lambda - const > 0$, then $S(t)$ is classical Compound Poisson process.
If $\mathbb{E}[\Lambda_t] = \mu_\Lambda < \infty$, then $\mathbb{E}[S_t] = a\mu_\Lambda$.
If $\mathbb{V}[\Lambda_t] = \sigma_\Lambda^2 < \infty$, then $\mathbb{V}[S_t] = (a^2 + \sigma^2)\mu_\Lambda + a^2\sigma_\Lambda^2$.
For more information on Cox and compound Cox processes see [10,11].

2.2   References, Descriptions of the Main Mathematical Results

Going back to the introductory part recall the main representation of the process of market price with stochastic time (2)

$$S\left(t\right) = \sum_{i=1}^{N_t} \Delta S\left(t_i\right) = \sum_{i=1}^{N_t} X_i,$$

where the counting process is a Cox process N_t, such that

$$\sum_{i=1}^{N_t} t_i \leq t < \sum_{i=1}^{N_t+1} t_i,$$

and $t_1, t_2, \ldots, t_{N_t}$ are the times of registered trades. The variables X_i are the price impacts of the corresponding trades.

We assume that $\{X_{t_i}, i = 0, 1, \ldots\}$ are i.i.d., $\mathbb{E}\left[X_i\right] = 0$, $\mathbb{V}\left[X_i\right] = \sigma_X^2 < \infty$.

Limit Theorem for Compound Cox Processes. In order to compare compound Cox process with other models of market prices we need to understand limit behavior of Cox processes when either intensity or time increases to infinity.

Let $d\left(t\right) > 0$ be a function growing unlimited when $t \longrightarrow \infty$.

Theorem 1. *Let $\Lambda_t \longrightarrow \infty$ in probability as $t \longrightarrow \infty$. For weak convergence to some random variable Z*

$$\frac{S\left(t\right)}{\sigma_X \sqrt{d\left(t\right)}} \Longrightarrow Z, t \longrightarrow \infty$$

it is necessary and sufficient that

1. $\mathbb{P}\{Z < z\} = \int_0^\infty \Phi\left(zy^{-\frac{1}{2}}\right) d\mathbb{P}\{U < u\} = \mathbb{E}\left[\Phi\left(zU^{-\frac{1}{2}}\right)\right], z \in \mathbb{R};$
2. $\frac{\Lambda_t}{d(t)} \Longrightarrow U, t \longrightarrow \infty.$

Condition 2 of the theorem means that intensity process Λ_t needs to grow in some regular way and, after normalization by $d\left(t\right)$, converge to some random variable U.

But it can remain stochastic.

It is also interesting that asymptotically distribution of $\frac{\Lambda_t}{d(t)}$ does not depend on t.

Condition 1 of the theorem generally speaking means that one-dimensional distribution of Cox process does not converge to Gaussian distribution.

Instead it converges to a mix of Gaussian distributions

$$\mathbb{E}\left[\Phi\left(zU^{-\frac{1}{2}}\right)\right]$$

where U is the mixing random variable.

Such mix can have a very heavy tailed distribution.

This explains why CLT does not hold in general case when time becomes random.

2.3 Dynamic VaR

Let $L(t)$ be the loss of a low latency strategy at time moment t, $\{\tau_1, \tau_2, \ldots\}$ – times of changes of $L(t)$ (events), $N(t)$ – number of events on $[0, t]$. As noted in [16], an important feature of actual transaction prices is the existence of noise (called microstructure noise in [16]). The process L can be decomposed within that framework into two parts

$$L(t) = \widetilde{L}(t) + S(t),$$

where $\widetilde{L}$ is semimartingale, S - noise. Consider the easiest case $\widetilde{L}(t) = \mathbb{E}L(t) = at$ $(t \geqslant 0)$. The linear trend term is predictable and sometimes can be compensated. We'll focus on the behaviour of noise.

Assume that $X_j = S(\tau_j) - S(\tau_{j-i})$ are i.i.d. normal random variables with zero mean (as a consequence of condition $\mathbb{E}S(t) = 0$) and $\mathbb{V}[X_j] = \sigma^2 < \infty$; $\{N(t), X_1, X_2, \ldots\}$ are independent; $N(t) = N_1(\Lambda(t))$, where $N_1(t)$ is homogeneous Poisson process with unit intensity; $\Lambda(t) = \int_0^t \lambda(s)\, ds$, instantaneous intensity $\lambda(s)$ is positive stochastic process with integrable trajectories; $\mathbb{E}[\Lambda(t)] = \mu t$.

Consider Cox process $S(t) = \sum_{j=1}^{N(t)} X_j$, $\bar{S}(T) = \max_{0 \leq t \leq T} S(t)$. We'll need the modification of the above theorem with fixed time interval and increasing flow intensity.

Theorem 2. [11, Theorem 2.2.1] Assume that $\Lambda(T) = \Lambda(T; \mu) \longrightarrow \infty$ in probability as $\mu \longrightarrow \infty$. Then

$$\mathbb{P}\left\{\frac{\bar{S}(T)}{\sigma\sqrt{\mu T}} < x\right\} \Longrightarrow F(x), (\mu \longrightarrow \infty)$$

iff there exists non-negative random variable U: $\frac{\Lambda(T; \mu)}{\mu T} \Longrightarrow U$, $\mu \longrightarrow \infty$. In addition $F(x) = 2\mathbb{E}\left[\Phi\left(\max(0, x)/\sqrt{U}\right)\right] - 1$.

The natural example of cumulated intensity process Λ (actually the random time change for the process N_1) is Gamma process. It is used in the popular Variance-Gamma model [12] as a subordinator for Brownian motion with drift.

Definition 4. The process $\{\gamma(t; \alpha, \beta) = \gamma(t), t \geqslant 0\}$ is called Gamma process with a shape parameter α and a scale parameter β iff

(1) $\gamma(0) = 0$ a.s.
(2) for any $0 \leqslant t_1 < t_2 < \ldots < t_n < \infty$ $\gamma(t_2) - \gamma(t_1), \gamma(t_3) - \gamma(t_2), \ldots, \gamma(t_n)$ $-\gamma(t_{n-1})$ are independent
(3) for any $t \geqslant 0, h > 0$ $\gamma(t + h) - \gamma(t)$ has gamma distribution with a shape parameter αh and a scale parameter β.

Substituting Gamma process as the random time change into $N_1(t)$ gives Levy process S. Then $\Lambda(T)$ has gamma distribution with a shape parameter αT and a scale parameter β. We can increase Poisson process N intensity by either increasing shape parameter or decreasing scale parameter. Choose the latter $\alpha = 1/T$, $\beta = (\mu T)^{-1}$. Then for any $x \geqslant 0$

$$\mathbb{P}\left\{\Lambda(T;\mu)/\mu T < x\right\} = 1 - \exp(-\beta x \mu T) = 1 - e^{-x}$$

and $\frac{\Lambda(T;\mu)}{\mu T}$ converges weakly to the exponentially distributed U with unit mean. Calculate $\mathbb{E}\left[\Phi\left(x/\sqrt{U}\right)\right]$ for $x \geqslant 0$

$$\mathbb{E}\left[\Phi\left(x/\sqrt{U}\right)\right] = \int_0^\infty \Phi\left(x/\sqrt{u}\right) d\left(1 - e^{-u}\right)$$

$$= \int_0^\infty \left[\frac{1}{2} + \frac{1}{\sqrt{2\pi}} \int_0^{x/\sqrt{u}} \exp\left(-\frac{v^2}{2}dv\right)\right] e^{-u} du.$$

Changing the order of integration gives

$$\mathbb{E}\left[\Phi\left(x/\sqrt{U}\right)\right] = 1 - \frac{1}{\sqrt{2\pi}} \int_0^\infty \exp\left(-\frac{u^2}{2} - \frac{x^2}{u^2}\right) du.$$

The last integral can be found in [9] formula 3.325. Finally

$$\mathbb{P}\left\{\frac{\overline{S}(T)}{\sigma\sqrt{\mu T}} < x\right\} \approx \begin{cases} 0, & x < 0, \\ 1 - e^{-\sqrt{2}x}, & x \geq 0. \end{cases} \tag{3}$$

From (3) we find q-level quantile of the maximum loss distribution and define Dynamic VaR as

$$D(T, q) = \frac{\sigma\sqrt{\mu T}}{\sqrt{2}} \ln \frac{1}{1-q}. \tag{4}$$

Note than the maximum loss distribution has heavier tail (exponential) than random variables X_j.

3   Example: Trading System Breakdown

3.1   Description of the Test Example

Dynamic VaR is used for real-time analysis of market risk associated with low latency strategies.

Here we reconstruct the situation similar to the incident of August 1, 2012 when reportedly a computer at Knight Capital Group (KCG Holdings Inc.) started mistakenly sending frequent orders buying shares at the ask level and immediately selling them at the bid level losing the bid-ask spread on each cycle.

As a result portfolio did not change more than 1 lot at any particular moment in time, but increased frequency of trades, 40 trades per second (or 25 ms between the trades), resulted in a steady rate of loss about \$10 million per minute.

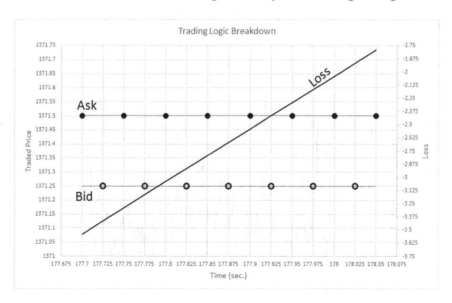

Fig. 1. Trading Logic Breakdown

The following chart explains our test case using S&P 500 e-mini futures (Fig. 1): by repeatedly buying the asset at the ask price and immediately selling it one tick lower at the bid level results in a significant loss even if only one contract is traded at a time.

The main reason for the loss in such case is not the size of the position and not a significant shift in the market price: in our example the level of the price is not changing. It is the frequency of trading that should alert risk management system as early as possible.

And what about the traditional measure for market risk, VaR, calculated from moving time window of 5 s?

It actually drops because the change of the Profit and Loss (P&L) becomes predictable which means that the standard deviation of the P&L increment drops down.

Since VaR is based on he standard deviation of the increment of the P&L it drops also.

This is shown on Fig. 2: during the period of software malfunction intensity of P&L changes jumps, at the same time standard deviation of the P&L drops down.

This, of course makes the traditional VaR not applicable to analysis of the market risk in low latency environment when risk has significant frequency component.

This explains why low latency trading systems do not use any of the traditional market risk measures in their safety logic.

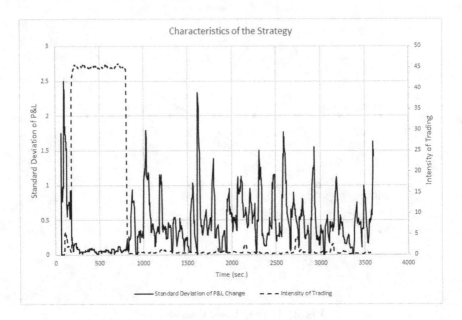

Fig. 2. Characteristics of the Strategy

Instead they use different ex-post measures such as realized P&L, i.e. the trading system would shut down when the realized P&L drops below the preset limit.

But would an appropriate predictive market risk measure give any benefit if used as part of the safety logic of trading robot?

To answer this question zoom into the previous picture to see events on the time scale at which high frequency trading systems make decisions.

On Fig. 3 we see the moment when the software accident occurs and the immediate aftermath of it.

3.2   Dynamic Var

We see on Fig. 3 the beginning of the drop of the standard deviation of P&L and simultaneous sharp increase in intensity of trading events.

Contrary to the traditional VaR that utilizes only standard deviation the new measure Dynamic VaR (DVaR) defined in the previous section (4) utilizes both characteristics: the standard deviation and the intensity.

On Fig. 4 we see that immediately after the incident DVaR (dashed line) jumps sharply to its maximum value in about 600 ms.

The realized P&L (dot-dashed line) used here as an ex-post measure of market risk has not reached the same level yet after 2 s.

In the environment when decisions are made in few microseconds advantage of the predictive measure DVaR relative to realized P&L can make a significant difference.

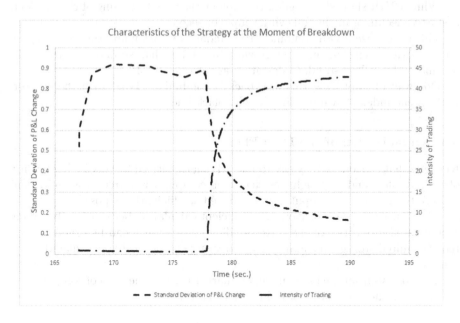

Fig. 3. Characteristics of the Strategy at the Moment of Breakdown

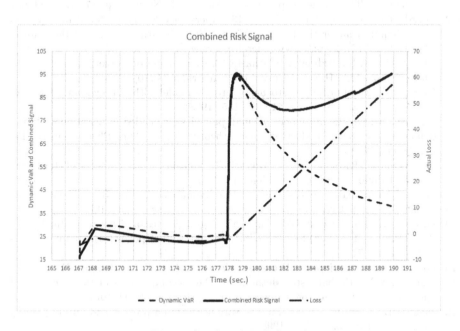

Fig. 4. Combined Risk Signal

While DVaR showed very good reaction at the very beginning of the incident it started declining in the following couple of seconds.

This is because DVaR combines the standard deviation and the frequency in a multiplicative expression: after growing in the first half second the product is dominated by falling standard deviation as seen on Fig. 3.

One way of improving this drawback is combining DVaR with the ex-post P&L in one index shown on Fig. 4 as the black line.

3.3 Organization of the Data Flow

Knowing that the low latency environment is characterized by very high intensity of the data flow and very tight restriction on computer resources it is important to have the process of computation organized as efficiently as possible.

The structure of DVaR allows very efficient calculation because it is based on accumulated intensities of the corresponding events which can be preprocessed, require limited storage in computer memory and can be easily recalculated when necessary.

This follows from the basic model for market price in the form of compounded Cox process and the results of the previous section.

References

1. Bachelier, L.: Theorie de la speculation. Ann. Ecole Norm. Sup. **17**, 21–86 (1900)
2. Board of Governors of the FRS, 2013 Supervisory Scenarios for Annual Stress Tests Required under the Dodd-Frank Act Stress Testing Rules and the Capital Plan Rule (2012). http://www.federalreserve.gov
3. Clark, P.K.: A Subordinated Stochastic Process Model of Cotton Futures Prices. Ph.D. Thesis. MA. Harvard University, Cambridge (1970)
4. Clark, P.K.: A subordinated stochastic process model with finite variance for speculative prices. Econometrica **41**, 135–155 (1973)
5. Einstein, A.: On the movement of small particles suspended in a stationary liquid demanded by by the molecular-kinetic theory of heat. Ann. Phys. (Leipzig) **17**, 549–560 (1905)
6. Embrechts, P., Klüppelberg, C., Mikosch, T.: Measuring Extremal Events for Insurance and Finance. Springer, Heidelberg (2003)
7. Fama, E.: Mandelbrot and stable paretian hypothesis. J. Bus. **36**, 420–429 (1963)
8. Fama, E.: The behavior of stock market prices. J. Bus. **38**, 34–105 (1965)
9. Gradshteyn, I.S., Ryzhik, I.M.: Table of integrals, series, and products. In: Jeffrey, A., Zwillinger, D., (eds.) Seventh edn. (Feb 2007) 1,171 pages ISBN number: 0-12-373637-4 (2007)
10. Grandel, J.: Mixed Poisson Processes. Chapman & Hall, London (1997)
11. Korolev, V.Y.: Probabilistic Methods for the Decomposition of the Volatility of Chaotic Processes. Moscow University Press, Moscow (2011). (in Russian)
12. Madan, D., Carr, P., Chang, P.: The variance gamma process and option pricing. Eur. Finan. Rev. **2**, 79–105 (1998)
13. Mandelbrot, B.B.: The variation of certain speculative prices. J. Bus. **36**, 394–419 (1963)

14. Mandelbrot, B.B.: The variation of some other speculative prices. J. Bus. **40**, 393–413 (1967)
15. Mykland, P.A., Zhang, L.: The double Gaussian approximation for high frequency data. Scand. J. Stat. **38**, 215–236 (2011)
16. Mykland, P.A., Zhang, L.: The econometrics of high frequency data. In: Kessler, M., Lindner, A., Sorensen, M. (eds.) Statistical Methods for Stochastic Differential Equations. Chapman and Hall, London (2012)

Generalized ATM Fraud Detection

Steffen Priesterjahn[1], Maik Anderka[2], Timo Klerx[3(✉)], and Uwe Mönks[4]

[1] DE R&D ACT 1, Wincor Nixdorf International GmbH, Paderborn, Germany
[2] DE R&D ACT 6, Wincor Nixdorf International GmbH, Paderborn, Germany
[3] Department of Computer Science, University of Paderborn, Paderborn, Germany
timok@mail.upb.de
[4] inIT – Institute Industrial IT, Ostwestfalen-Lippe University of Applied Sciences,
Lemgo, Germany

Abstract. Recent activities in attacks on automated teller machines have shown a sophistication that has grown to a degree, where it is not always technically possible to prevent the attack. This paper describes an approach for anomaly and attack detection for ATMs. The approach works on multiple levels. First, we use sensor fusion on the low-level hardware sensors to get robust information about the device state. Second, we use a new model-based and self-learning anomaly detection method on the diagnosis data of all ATM devices to robustly detect anomalies in the system that might indicate an attack on the machine.

1 Introduction

Since their invention in the early 1970s, Automated Teller Machines (ATM) have been subject to multiple types of attacks. In their early years those attacks were typically targeted at the cash inside the machine. In later years, this was supplemented by attacks on the private data of the ATM users, namely the card data and the user's personal identification number (PIN). Recent years have seen a substantial increase both in number and sophistication of ATM attacks (detailed in Sect. 2). This can be seen both in the official statistics of the relevant police forces [3], but even more in the statistics that are internally available to banks and ATM manufacturers. Due to the importance of ATMs in the overall cash flow and the global cash supply, fighting this kind of crime has become a major interest for both the banking industry and the police.

Yet, similar to the recent developments in the anti-virus industry [23], it has become harder and harder to prevent all types of attacks by security technology. Most of today's security solutions are highly specialized countermeasures against certain attacks and represent answers to ever-changing attack techniques that have been found in the field over the years. However, the industry is currently facing a new level of sophistication on the attacks that makes it harder and harder to find technical solutions that can actually prevent attacks without sacrificing the serviceability of the ATMs. In recent years, the attackers have moved from small solitary groups to multinational organized crime organizations with considerable development and reverse engineering capabilities that are able to

© Springer International Publishing Switzerland 2015
P. Perner (Ed.): ICDM 2015, LNAI 9165, pp. 166–181, 2015.
DOI: 10.1007/978-3-319-20910-4_13

attack large numbers of ATMs in a short time. Therefore, attack detection and mitigation technologies are moving more and more into focus: If it is not possible to prevent an attack, the ATM should at least be able to detect that something strange is going on and act accordingly by shutting itself down or providing an alarm signal.

This paper provides an overview of the recent research activities of Wincor Nixdorf and its partner universities, University of Paderborn and Ostwestfalen-Lippe University, in the field of generic anomaly detection in technical systems. This research will be the basis of the currently developed attack detection systems for Wincor Nixdorf ATMs and has the goal to detect possible attacks by the intelligent combination of all available sensor information in the ATM. The joint project is funded by the German Federal Ministry of Research and Education (BMBF) as the funding project itsowl-InverSa[1], which belongs to the Leading-Edge Cluster it's OWL[2].

The paper is structured as follows. First, we give an introduction into ATM security and the internal architecture of an ATM. Then we describe, how we use sensor fusion on the low level sensors to gain robust and interesting information that is then analyzed by the InverSa anomaly detection system, as described in Sect. 5. We then present an experimental analysis of the approach before we conclude the paper.

2   ATM Fraud and Current Security Technology

This section provides an overview of the most prominent ATM attacks and countermeasures. See Figs. 1 and 2 for reference.

Skimming. Skimming attacks target the data that is stored on the card of the ATM user to later use it to withdraw money. Usually, the main object of interest is still the magnetic stripe data that typically holds the user's banking account data and some other additional information.[3] A typical card skimmer consists of a very small magnetic read head and some devices to either store or transmit the captured data. They are positioned in front or inside the front area of the card reader of the ATM. Most recently, there have also been findings of chip-based skimming devices that are placed between the smart card reader and the card.

Card Trapping. The objective of card trapping attacks is to capture the card instead of just the data. Typical attacks use small loops or other devices to jam the card in the card reader. When the user leaves the ATM to inform the bank manager about the jammed card, the attacker quickly uses a corresponding device to retrieve the jammed card.

PIN Capturing. To successfully withdraw money from a bank account, the attacker not only needs the card data, but also the PIN of the users. Therefore, each skimmer is usually accompanied by a second device to capture PIN

---

[1] German project title: "Intelligente vernetzte Systeme für automatisierte Geld-kreisläufe".

[2] Intelligent Technical Systems OstWestfalenLippe: http://www.its-owl.de.

[3] Due to legacy support, this information is usually unencrypted.

entries. This is very often done by a miniature camera behind a hidden panel or by PIN pad overlays. A PIN pad overlay mimics the surface of the original PIN pad and which is put on top of the original PIN pad. It registers the pressed keys and stores the information in a recording device.

Cash Trapping. Cash trappers are devices that are put in front or inside the cash output slot of the ATM. They trap or capture the dispensed money and make it inaccessible to the user. When the user then leaves the ATM, the attacker quickly captures the money from the manipulated ATM.

Forced Opening. A wide range of ATM attacks just try to forcibly open the safe that holds the money. The most typical attack in this area is to fill the dispenser with some gas that is then brought to explosion. Other attacks use the cash retract function—which retracts the money into the ATM if it is not taken—to place a flat sheet of plastic explosives inside the safe.

Malware. In recent years the amount of software-based attacks has risen considerably. The typical target of this attack is to gain software access to the cash dispenser to issue cash dispense commands and to empty the ATM. However, there are numerous other attack scenarios that have been encountered in the field, e.g. capturing user data, modifying the ATM-to-host communication or simply to bring the ATM out of service. To protect the ATM against such attacks, the ATM manufacturers have come up with numerous solutions and products to, most preferably, make the attack impossible or to at least detect the attack.

Securing the Safe. Recent technology trends are safes that even survive a gas explosion by discharging the internal pressure in a controlled way. The safe locks have become more secure as well, e.g., with code locks, time-based locks, etc. In addition, many ATMs now have gas sensors that raise an alarm, when the concentration of explosive gas in the safe has reached a threshold.

Securing the PIN Pad. The PIN transport is always encrypted. In most countries, each PIN is encrypted by a uniquely derived key and can only be decrypted at the banking host. Therefore, it is impossible to capture the PIN with a software-based attack. In addition, the PIN pad itself is a high security device that destroys its keys, if it is opened or removed from the system. Protecting the PIN pad from PIN capturing attacks, however, is a different matter. There currently exists only one technical solution to detect PIN pad overlays and no solution at all to protect ATMs against PIN capturing via camera.

Securing the Card Reader. Nowadays the security of the card reader relies on the secure communication between the smart card chip and the host. However, there still exist several countries, where card security is still based on the magnetic stripe or uses it as the fallback solution. To secure the card reader against skimming attacks, there exist a plethora of solutions:improved mouth pieces to impede easy skimming, intelligent mouth pieces that generate an alarm when removed, electro-magnetic skimming detectors (based on metal detection), optical skimming detectors (based on cameras), or electromagnetic field generators to prevent skimming at the ATM.

Fig. 1. Internal components of a Wincor Nixdorf CINEO 4060 Cash Recycling Machine.

Fig. 2. Example for a Wincor Nixdorf ATM Fascia.

Securing the Cash Dispenser. There exist several approaches and solutions against cash trapping attacks, ranging from cash dispense slot design changes to optical or camera-based solutions. In addition, the communication to the cash dispenser is encrypted in most countries. However, only very few banks have an end-to-end encryption in place to ease servicing. Recent years have seen a constant increase in software attacks on the cash dispenser ranging from direct attacks against cash dispenser firmware to attacks on all software layers of the ATM.

Securing the Software. ATM software security solutions typically try to prevent access to important system resources and to prevent malware from attacking the system. Today's software security suites typically contain three sub-components: operating system hardening, hard disk encryption, and a host intrusion prevention system (HIPS). In contrast to typical anti-virus solutions, which allow everything to run that is not explicitly forbidden, HIPS generally only allows those operations on the system that have been explicitly configured.

3   The Architecture of an ATM

ATMs are very complex mechatronic systems that consist of many independent and sophisticated sub-components and devices. Figure 1 shows the internal components of a state-of-the-art cash recycling machine. It consists of a cash recycling module with eight cassettes, an encrypted PIN Pad, a motorized card reader, a coin module, a "17" touch screen, a journal printer, special electronics, and a customized PC. This is only one variant of an ATM. Other variants can include different types of safes, PCs, a simple cash dispenser, a check scanner

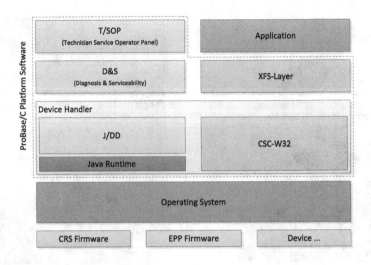

Fig. 3. The Wincor Nixdorf software stack.

(single or bundle), a non-motorized card reader, a contactless card reader, a second monitor and an operator panel, a receipt printer, a barcode reader, digital cameras, etc.

Each of these devices has to be able to operate with a high availability and robustness. ATMs are usually operated for at least seven years. Several banks even have replacement cycles that are at about ten to twelve years or longer. All the devices in the ATM have to be serviceable and secure.

Many of the internal ATM devices are either mechanically complex or have to be secured against possible attacks. Therefore, they usually contain a high number of low level sensors. For example, light barriers determine the current position of a card in the card reader or the position of a banknote on its way in the cash dispenser, Hall effect sensors are used to measure the banknote thickness in the cash dispenser, and imaging devices are applied in cash recyclers, scanners, barcode readers, or security cameras. One of the targets of this project is to use sensor fusion on these low-level sensors, combined with cheap additional sensors, to gain robust information about the operation state of the respective devices. Section 4 overviews of the corresponding approaches.

Most devices in the ATM have own computing abilities and can act rather autonomously via their firmware. However, at the end, they are all connected to the ATM-PC that is controlling the whole ATM and providing the user interface. The standard for device communication in ATMs is currently the Universal Serial Bus (USB).

The PCs themselves are highly customized with additional connectors and are usually run with Microsoft Windows. Windows has been the prominent ATM operating system since the mid-90s. Before that, MS-DOS and OS/2 were used. The ATM operators—i.e., banks—usually prefer Microsoft Windows to any other operating system because it allows them to include the ATMs into their default IT infrastructure. Several attempts to use other operating systems from ATM vendors have failed in the past.

Figure 3 shows a typical software stack on a Wincor Nixdorf ATM. The operating system is supplemented by additional device drivers and the platform software. For all major ATM vendors, the platform software is a runtime component, that takes control over the ATM devices and provides a common interface to a possible ATM application. The ATM application is the program, that is basically running the ATM and providing the user interface. The interface between the ATM application and the platform software has been standardized in eXtensions for Financial Services (XFS)[4] by the Comité Européen de Normalisation (CEN).

The platform software of an ATM usually contains a diagnosis and serviceability (D&S) module that monitors the ATM health state and helps to solve possible problems in the ATM. The D&S module is the single instance in the ATM where all device information is handled. Therefore, it is the basis of the anomaly detection approach described in Sect. 5. Listing 1 shows an example event log from the D&S module of a Wincor Nixdorf ATM. It contains events from the special electronics (SEL), the encrypted PIN Pad (EPP), the journal printer (PRT), and the card handling device (CHD). Each event contains a certain message ID that provides information about the type of event. Some events also contain additional values, that provide extended status information.

Listing 1. Excerpt from the event log of a Wincor Nixdorf ATM.

```
17.06.2013  11:37:13.968  [SEL]  MessageId: SE_DOOR_OPEN
17.06.2013  11:37:14.171  [EPP]  MessageId: EPP_SERIAL_STATUS , Value: FFTEST1023101407
17.06.2013  11:37:14.171  [EPP]  MessageId: EDM_REMOVAL_SWITCH_STATUS , Value: 192
17.06.2013  11:37:52.921  [PRT]  MessageId: TP07_SENSOR_CODE , Value: 0
17.06.2013  11:37:53.000  [PRT]  MessageId: TP07_ERROR_CODE , Value: 0
17.06.2013  11:37:57.390  [CHD]  MessageId: CHD_INIT_START_STATUS , Value: 2000000000000
17.06.2013  11:37:57.390  [CHD]  MessageId: CHD_TYPE, Value: 50
17.06.2013  11:38:00.343  [CHD]  MessageId: CHD_INIT_STOP_STATUS , Value: 2000000000000
17.06.2013  11:38:00.343  [CHD]  MessageId: CHD_STATUS_TRANSPORT , Value: 0
```

4   Sensor Information and Sensor Fusion

As described in Sect. 3, an ATM consists of several devices containing dedicated sensors necessary for operating both the single device as well as the whole ATM. Therefore, the sensors capture the current state of a physical quantity, which is necessarily used in the device's control for both safe and secure operation. The idea is to use these already available solitary sensors, especially their acquired signals, and bring them together by concurrent processing—hence fuse them. This will lead to a broader view on each device to monitor its condition on a low level close to the hardware: By processing single signals, certain effects can be detected, but might also lead to wrong decisions regarding the current situation. Such problems are omitted by taking additional information from other signal sources acquiring the same situation into the decision process. The following thought experiment shall clarify this: consider the detection of the magnetic field by a Hall effect sensor attached to the card reader. The system will likely infer that a skimming device has been attached to the ATM. Instead, a customer's crutches have been causing the change in the magnetic field. This situation can

---

[4] http://www.cen.eu/work/areas/ict/ebusiness/pages/ws-xfs.aspx.

be clarified by, e.g., involving images from cameras attached to the ATM, which certainly detect crutches instead of a skimming device.

The aforementioned example illustrates two challenges to be tackled: The necessity for the combination (or fusion) of several signal sources, which are likely of different types, and the need to cope with conflicting statements derived from single sources.

Approaches to sensor fusion can be divided into three main categories [7]:

- *Data level fusion* involves combining raw sensor signals prior to performing any data transformations. In order to combine sensor signals at the data level, the signals must originate from sources which produce the same type of signal.
- *Feature level fusion* involves first extracting features, which describe the data, and then combining the features from each signal to produce a fused signal. Unlike data level fusion, feature level fusion can be applied to data from both homogeneous and heterogeneous sensor types.
- *Decision level fusion* involves making a local decision from each signal and then combining the decisions to get the final output. Another approach to decision level fusion is to apply a mining algorithm that determines the probability of each possible outcome from each signal, and then combines the probability of class membership for each class using summing or voting [20].

An approach followed usually is the individual evaluation of the signals on *data level*. One simple possibility is to check if each signal's amplitude is within predefined boundaries. An anomaly would be detected, if one signal was outside them (cf. the previously given example). This simple approach fails for complex systems like ATMs simply due to the sheer amount of signals, which are to be processed. Additionally, this model is just too simple to adequately represent the ground truth. Instead, we carry out fusion on the *feature level*, which is suitable to include also *decision level* information.

Our preferred fusion model is described in the following: As physical signals are always prone to imprecision and uncertainty, representation concepts must be capable of modeling these. Due to its comprehensive theoretical foundation, probability theoretical methods like Bayesian fusion naturally seem to be an adequate candidate to represent the acquired data [4]. Uncertainties in ATM monitoring (and machine condition monitoring in general) are mainly of epistemic nature, but probabilistic methods are only able to model aleatoric (random) uncertainties appropriately [14]. Besides, Kolmogorov's axiomatic foundation of probability theory contains the additivity axiom [11], which is impracticable in real-world applications: In case a hypothesis (e.g., "The ATM is functioning correctly") is denied, this axiom implicitly leads to the acceptance of its complement [8], hence the machine's incorrect function. This drastic view is incorrect in many cases, everything in between is true instead (e.g., a part of the card reader is worn out and not working properly, but is still working).

As already mentioned before, conflict between data sources must be considered appropriately in the fusion process. Bayesian fusion is inappropriate here, as well [2]. Instead, the representation of the features extracted from the acquired data by fuzzy sets [24] is considered beneficial, as they are capable of modelling epistemic uncertainty properly, and expert knowledge can be included

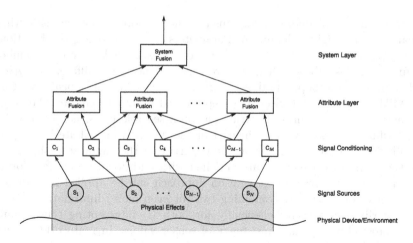

Fig. 4. *Multilayer attribute-based conflict-reducing observation* system MACRO (cf. [17]).

straightforward [14]. The subsequent combination or fusion for decision making, respectively, of the information is naturally carried out suitably by fuzzy logic methods, but these intrinsically have no built-in conflict resolution mechanisms. Instead, Dempster's rule of combination [21] is capable to handle such cases. This fusion approach does not imply the acceptance of an hypothesis' complement in the case of its denial [8], but it leads to counter-intuitive results in high-conflict situations [18]. Due to this, an alternative approach, which is based on Dempster's rule of combination and tackling its problems, has been defined, extended, and refined. This has led to the *fuzzified balanced two-layer conflict solving* (μBalTLCS) aggregation and combines the conflict solving capabilities of Dempster's rule of combination with fuzzy set's and fuzzy logic's uncertainty handling [18].

The novel fusion approach is a core part of the *multilayer attribute-based conflict-reducing observation* system MACRO depicted in Fig. 4. It is an information fusion structure for machine condition monitoring applications, described and analysed in detail in [14,15,18]. MACRO's fusion procedure is as follows: For determining the state of a complete system, signals of the system itself as well as its environment are acquired by sensors *(signal sources)*. The previously mentioned features are extracted in the following *signal conditioning* step which may also include signal pre-processing procedures. A unimodal potential function—already widely used in many industrial monitoring applications—is used as the prototype fuzzy membership function modeling the normal state of each input signal [13]:

$$\mu\left(x,\mathbf{p}\right) = \begin{cases} 2^{-\left(\frac{|x-S|}{C_l}\right)^{D_l}}, & x \leq S \\ 2^{-\left(\frac{|x-S|}{C_r}\right)^{D_r}}, & x > S. \end{cases}$$

The function's shape parameters $\mathbf{p} = \left(S, C_{l/r}, D_{l/r}\right)$ are trained semi-automatically based on actual data acquired from the monitored ATM. Width parameters $C_{l/r}$ are trained as described in [13]. The function's center of gravity S

is every feature x's median to make this parameter more robust against outliers compared to its original definition. Parameters $D_{l/r}$ controlling the function's edge steepness are user-defined. Each feature x is modelled by one membership function. Ensembles of features are then grouped to k different *attributes* which represent the supervised ATM in its actual physical/logical structure: each ATM component (cf. Sect. 3) is represented by one attribute. The actual combination of the features is carried out on a sensible basis, hence such that each attribute represents a certain part of the ATM. An attribute is compiled manually in most cases, relying on the expertise of the engineer/designer creating the attributes. If signals are enriched with additional semantic information, an automatic attribute generation approach can be applied instead [16]. Redundancies occurring by combining two (or even more) information sources to one attribute are used beneficially for both (i) intercepting sensor faults and (ii) cross-checking the consistency of sensor values. In real-world applications, it is often sufficient to monitor the *attribute health*, i.e., to which degree $^Nm(A_k) = {}^N\mu(A_k)$ the attribute A_k is in normal condition ^NC. The necessary counter-proposition "abnormal condition" ^AC is modeled by the dual measure $^Am(A_k) = {}^A\mu(A_k) = 1 - {}^N\mu(A_k)$ and can thus be substituted by $^Nm(A_k)$.

Information of n sources is fused by μBalTLCS [18] on the *attribute layer*. This fusion approach combines every two signal sources' data so that a conflict is considered and solved between them. The conflict is further resolved at a *local group level* by additive combination of non-conflicting ($^Nm_{nc}(A_k)$) and conflicting parts ($^Nm_c(A_k)$): $^Nm(A_k) = {}^Nm_{nc}(A_k) + {}^Nm_c(A_k)$, where the non-conflicting mass is:

$$^Nm_{nc}(A_k) = \frac{2}{n \cdot (n-1)} \cdot \sum_{i=1}^{n-1} \sum_{j=i+1}^{n} {}^Nm_i \cdot {}^Nm_j,$$

and the conflicting mass is:

$$^Nm_c(A_k) = k'_{cm}(A_k) \cdot \frac{1}{n} \sum_{i=1}^{n} {}^Nm_i, \quad k'_{cm}(A_k) = \frac{2}{n} \sum_{i=1}^{n} {}^Nm_i - 2 \cdot {}^Nm_{nc}(A_k).$$

The conflicting coefficient $k'_{cm}(A_k) \in [0,1]$ represents the amount of conflict inside an attribute between the associated information sources. It regulates an attribute's importance $I_m(A_k) = 1 - k'_{cm}(A_k)$: The higher the conflict, the smaller its importance.

Importance of an attribute plays a major role in the fusion on *system layer*. Here, we use the *Implicative Importance Weighted Ordered Weighted Averaging* (IIWOWA) class of fuzzy averaging operators [12] to obtain the overall *system health* ^Nm. Each attribute health is at this step weighted with its corresponding importance so that attributes bearing high conflict have only small influence on the system health. Hence, MACRO allows to determine individual device conditions on attribute level as well as the state of the complete ATM on system layer. Due to this resemblance a transparent supervision system is created that helps interpreting results during the ATM's runtime.

Additionally, MACRO is beneficial in decreasing the complexity in the creation of sensor and information fusion systems. We have already shown that by enriching the applied signal sources with additional self-descriptive information (e.g., the component a sensor is belonging to; measurement quantity; measurement unit), this information may be used for automatic attribute generation [16,17]. During the ATM production process, different configurations are considered accordingly without manual input to setup the monitoring system automatically. Together with communication self-configuration methods, a step towards plug-and-play in the industrial environment is taken.

Our sensor fusion model MACRO creates robust and dense information of higher quality from the low level sensor signals. This information is suitable to be interpreted by the D&S module and is transferred to the next higher processing level, the InverSa anomaly detection system, which is described in the following section.

5   The InverSa Anomaly Detection System

We tackle ATM fraud detection as a *model-based anomaly detection* problem.[5] Model-based anomaly detection, also known as model-based diagnosis, is an important application field of artificial intelligence. It deals with the algorithmic analysis of whether a system operates abnormally given a system description (a model) and observations of its behavior. The fundamental procedure of model-based anomaly detection is as follows: During operation, the model is used to simulate the system's behavior, and if the expected behavior predicted by the simulation significantly differs from the observed behavior of the real system, an anomaly has occurred. In the ATM context, significant anomalies can be assumed as strong indicators of fraud or manipulation.

The InverSa anomaly detection system is a two-step approach, which comprises a model learning phase and an anomaly detection phase:

1. *Model learning.* In this phase, a behavior model is automatically created by means of machine learning techniques. This requires a certain amount of observations that reflect the normal behavior of an ATM.
2. *Anomaly detection.* In this phase, the ATM's real-time behavior is monitored, and in case of detecting abnormal behavior with respect to the learned model, a security alert is triggered.

The following subsections describe the model learning phase (Sect. 5.1) and the anomaly detection phase (Sect. 5.2). Moreover, we report on an empirical evaluation study showing the practical applicability of our anomaly detection system (Sect. 5.3).

5.1   Learning ATM Behavior Models

In order to learn an appropriate behavior model, it is necessary to define the crucial system properties of an ATM. Following [9], an ATM can be interpreted as a *discrete event system*. Discrete event systems are characterized by three basic properties [6]:

---

[5] This approach has been proposed in [9] and further refined in [10].

1. *Discrete states.* The system's state space is finite.
2. *Dynamic.* Each new state of the system depends on the state's predecessors.
3. *Event-driven.* State changes are caused by asynchronously occurring events.

Moreover, in practice, discrete event systems frequently operate in a stochastic setting, i.e., a system state describes a random process, such as unpredictable effects of nature or user interactions. Taking the event logs produced by an ATM's Diagnosis and Serviceability platform (described in Sect. 3) as a basis, an ATM can be interpreted as a discrete event system: The events occur at various time instances (event-driven), the system's state space is finite since no continuous variables are considered (discrete states), and the system states are interdependent because ATMs typically operate in a transaction-based manner (dynamic). Moreover, stochastic elements are brought in by customer interactions and environmental conditions.

Klerx et al. [10] proposed a tailored stochastic timed model for discrete event systems, namely *probabilistic deterministic timed-transition automaton* (PDTTA). Roughly speaking, a PDTTA is a traditional probabilistic deterministic automaton with additional time probability distributions at the transitions. Figure 5 shows an example of a PDTTA. Formally, a PDTTA A is defined as follows:

Definition 1 (PDTTA). *A probabilistic deterministic timed-transition automaton, denoted by A, is a six-tuple $A = (S, s_o, \Sigma, T, \xi, \tau)$, where*

- *S is a finite set of states, with $s_0 \in S$ the start state.*
- *Σ is a finite alphabet comprising all relevant events (status messages).*
- *$T \subseteq S \times \Sigma \times S$ is a finite set of transitions. For example, $\langle s, e, s' \rangle \in T$ is the transition between state $s \in S$ and state $s' \in S$ triggered by event $e \in \Sigma$.*
- *$\xi : T \to [0, 1]$ is a transition probability function, which assigns a probability value p to every transition.*
- *$\tau : T \to \Theta$ is a transaction time probability function, which assigns a probability distribution $\theta \in \Theta$ to every transition, where Θ is a set of probability distributions. Every $\theta \in \Theta$ has the signature $\theta : \mathbb{I} \to [0, 1]$, where $\mathbb{I} \subseteq \mathbb{N}$ is a set of time values.*

Learning a PDTTA from (normal) observations of an ATM provides us an adequate behavior model for anomaly detection. While the learning of simple (untimed) model types can be considered as being state of the art, the learning of probabilistic deterministic timed automata is still a challenging task (see, e.g., [19]). We therefore apply a two-step approach for PDTTA learning.

In the first step, a traditional probabilistic deterministic finite automaton (PDFA) is learned using a state-of-the-art approach.[6] The PDFA models the ATM's sequential behavior, without considering any time information. As

---

[6] A PDFA corresponds to a PDTTA without transaction time probability function τ (cf. Definition 1). For further information about probabilistic (timed) automata, refer to [1].

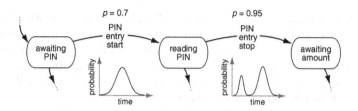

Fig. 5. Example of a probabilistic deterministic timed-transition automaton (PDTTA). For each transition, a probability p for taking the transition is given as well as a probability distribution describing the likelihood for the event's timing, e.g., the time it takes to enter the PIN. (Source: [10])

already mentioned, learning a PDFA from sequence data is a well-studied problem, and several approaches have been proposed to solve this task (see, e.g., [22]). We apply the ALERGIA algorithm, which is one of the most popular PDFA learning methods (for further information, refer to [5]).

In the second step, the PDFA is augmented with the relevant timing information to obtain the final PDTTA. Specifically, this means learning the function τ (cf. Definition 1). We therefore iterate over the input observations and traverse the initial PDFA. For each transition, we store the time values that belong to the events that triggered the transition. After processing all input observations, we construct a probability distribution θ for each transition by fitting a probability density function to the stored time values. Fitting is performed using Kernel Density Estimators with Gaussian Kernel.

5.2 Detecting Abnormal Behavior

We will now describe how the learned model can be used to detect abnormal behavior. More specifically, in the context of ATM fraud detection, the goal is to detect abnormal behavior that results from attacks or manipulation (cf. Sect. 2). However, certain kinds of attacks result in different types of abnormal behavior. Klerx et al. [10] identify four general anomaly types that occur in discrete event systems: anomalous event, anomalous event sequence, anomalous event timing, and anomalous event sequence timing. In the context of ATM fraud detection, an example of an anomalous event could be a cash dispense event without preceding card entry and PIN entry events. This might indicate a software-based attack, where the attackers have gained access to the cash dispenser to issue cash dispense commands. An example of an anomalous event sequence could be caused by a manipulation of the software or the communication with the banking host, where a certain unusual event sequence is used as a trigger to dispense higher amounts of cash or to circumvent the PIN check. An anomalous event timing could be caused if the PIN entry takes an unusual long time. This might be due to a PIN pad overlay, which delays the PIN entry process. Finally, an anomalous event sequence timing could be caused by a software or denial-of-service attack that leads to a general slowdown of the overall system.

Our model-based anomaly detection approach is able to identify all four anomaly types based on a PDTTA. In particular, the anomaly detection task can be stated as follows: Given a learned PDTTA A and an observation X, decide whether X is normal or abnormal with respect to A. In a real-world scenario, X could be a sequence of log events from the event logs of an ATM in operation (cf. Sect. 3). An event sequence typically corresponds to a single transaction, which starts with a "card entry" event and ends with a "card out" event. We apply a two-step approach for anomaly detection.

In the first step, we gather the event and time probabilities associated with the given observation when traversing the trained model. We start in the initial state of the model. From the current state we traverse the transition associated with the observed symbol while storing the probability for this transition and the probability for the observed time value for the taken transition. Then, the new current state is the target state of the transition we took. We repeat this procedure until the observation ends.

In the second step, we aggregate the event and time probabilities separately and decide on the aggregated values whether the observation is an anomaly. We therefore separately aggregate all gathered event and time probabilities by multiplying them with each other and normalizing the result according to the length of the observation. This gives us two anomaly scores: one for the aggregated event probabilities (a_E) and one for the aggregated time probabilities (a_T). Each of the anomaly scores is then compared to a predefined threshold (c_E, c_T). If one of the anomaly scores a_X is lower than the corresponding threshold c_X we state that the observation is an anomaly.

5.3   Empirical Evaluation

The goal of the evaluation is to assess the anomaly detection effectiveness of our approach. We therefore recorded event logs from a public ATM in a period of nine months. The resulting log file has a size of 1.6 GB and is structured as shown in Listing 1. The file contains 15 million status messages (events). In a preprocessing step, we split the log file at "card out" events, as mentioned above, in order to identify individual transactions. Each transaction represents a single observation. Moreover, to account for seasonality effects, we separate the data into weekly chunks. Subsequent chunks are used as training, validation and test set. The training set is used for model learning. The validation set is used to find the optimal combination of parameters and thresholds. Finally, the test set is used to assess the performance of our approach with the chosen parameter combination on an unseen set of observations. The anomaly detection effectiveness is measured in terms of precision, recall, and F-measure. Precision is the ratio between correctly detected anomalies and all anomalies. Recall is the ratio between detected anomalies and all anomalies. F-measure is the harmonic mean of precision and recall.

In order to evaluate the anomaly detection effectiveness, data that contains both normal *and* abnormal observations is required. The recorded event logs, however, contain only normal observations. Note that data of attacks is in general not available for reasons of security and secrecy. To enable a proper evaluation, anomalies can be created in different ways [10]: by simulating physical attacks in a lab, by a software-based simulation of attacks using a (learned) behavior

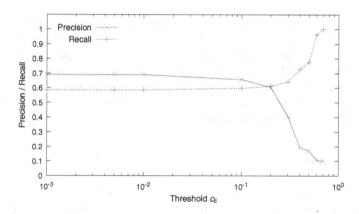

Fig. 6. Anomaly detection effectiveness of a PDTTA in terms of precision and recall over the event threshold c_E (x-axis in log scale). The time threshold c_T is fixed, $c_T = 10^{-5}$. (The results for varying threshold c_T while fixing threshold c_E are similar.)

model, and by generating artificial anomalies using statistical methods. Physical simulation is complicated because a real system has to be modified or even damaged, which is expensive and therefore inefficient. Software-based simulation requires both an interpretable behavior model as well as a human domain expert to alter the model accordingly, which is complicated and not always possible. We hence generated artificially anomalies by modifying randomly chosen observations from the event logs in order to derive abnormal observation that correspond to the four anomaly types described above (for details refer to [10]).

Given a test set where the ratio of artificial anomalies is 10 %, our approach achieves an F-Measure value of 0.63. Although the absolute value seems to be rather low, our approach clearly outperform a baseline approach that uses a PDFA (without any timing information) and that achieves an F-Measure value of 0.51. This result shows that the timing information in the PDTTA creates additional value, which increases the anomaly detection effectiveness. We also investigated the effectiveness for each of the four anomaly types. Whereas the results for anomalous events and anomalous event timing fail to come up to expectations, the results for anomalous event sequences and anomalous event sequence timing are quite promising (F-Measure values of 0.9 and 0.71 respectively). The discrepancy between event-based anomalies and sequence-based anomalies is one topic of our current work, which also involves increasing the overall performance.

Another feature of our approach is the possibility to adjust the anomaly detection effectiveness to the needs of users or customers. By varying the thresholds (c_E and c_T) the precision/recall tradeoff can be controlled. This is shown exemplarily in Fig. 6 for the event threshold (c_E). For higher values of c_E more anomalies are detected (increasing recall), but also more normal sequences are falsely labeled as anomaly (decreasing precision). Controlling this tradeoff is important to adjust the sensitivity of the approach; depending on whether the anomaly detection should be as complete as possible or as accurate as possible, i.e., detecting virtually all anomalies and accepting a large number of false alarms versus preferably preventing false alarms and possibly not detecting all anomalies. Figure 6 also shows that for $c_{\mathbb{E}} = 10^{-2}$ or lower both precision and recall stay almost constant. In the current approach it is not possible to tune the precision higher than 0.7 (this is also part of our current research). Nevertheless, the results show that this is the right course of action, and that our system is already able to detect certain types of anomalies with a quite good precision while maintaining a reasonable recall.

6 Conclusion

The presented results show that the described approaches can form the basis for the possible implementation of a robust anomaly detection system. By using a layered approach, where the MACRO approach is used to fuse low-level sensor signals to obtain robust and dense high-quality information, and where this information can later be used in a higher processing level for anomaly detection. Our model-based anomaly detection approach automatically learns a behavior model from ATM event logs, which is then used during operation to identify abnormal behavior that indicates fraud or manipulation. The experimental results shows, that the selected PDTTA-model can outperform state-of-the-art anomaly detection methods. The next steps towards a fully functional anomaly detection solution are currently being performed by the integration of additional sensors and by extending the anomaly detection approach with different approaches to increase robustness and detection accuracy.

Acknowledgements. This work was supported by the Wincor Nixdorf International GmbH, and partly funded by the German Federal Ministry of Education and Research (BMBF) within the Leading-Edge Cluster "Intelligent Technical Systems OstWestfalenLippe" (it's OWL).

References

1. Alur, R., Dill, D.L.: A theory of timed automata. Theoret. Comput. Sci. **126**(2), 183–235 (1994)
2. Ayyub, B.M., Klir, G.J.: Uncertainty Modeling and Analysis in Engineering and the Sciences. Chapman and Hall/CRC, Boca Raton (2006)
3. Bundeskriminalamt: Polizeiliche Kriminalstatistik 2013. Bundeskriminalamt, Kriminalistisches Institut, Fachbereich KI 12, 65173 Wiesbaden (2014)
4. Carl, J.W.: Contrasting approaches to combine evidence. In: Handbook of Multisensor Data Fusion, pp. 7-1-7-32. CRC Press (2001)
5. Carrasco, R.C., Oncina, J.: Learning stochastic regular grammars by means of a state merging method. In: Carrasco, R.C., Oncina, J. (eds.) ICGI 1994. LNCS, vol. 862, pp. 139–152. Springer, Heidelberg (1994)
6. Cassandras, C.G., Lafortune, S.: Introduction to Discrete Event Systems. Springer, Heidelberg (2008)
7. Hall, D.L., Llinas, J.: An introduction to multisensor data fusion. Proc. IEEE **85**(1), 6–23 (1997)

8. Jousselme, A.L., Maupin, P., Bossé, E.: Quantitative approaches. In: Concepts, Models, and Tools for Information Fusion, pp. 169–210. Artech House (2007)
9. Klerx, T., Anderka, M., Kleine Büning, H.: On the usage of behavior models to detect ATM fraud. In: Proceedings of the 21st European Conference on Artificial Intelligence (ECAI 2014). pp. 1045–1046. IOS Press (2014)
10. Klerx, T., Anderka, M., Kleine Büning, H., Priesterjahn, S.: Model-based anomaly detection for discrete event systems. In: Proceedings of the 26th IEEE International Conference on Tools with Artificial Intelligence (ICTAI 2014). pp. 665–672. IEEE (2014)
11. Kolmogorov, A.N.: Foundations of the Theory of Probability. Chelsea Publishing, New York (1950)
12. Larsen, H.L.: Efficient importance weighted aggregation between min and max. In: Proceedings of the 9th International Conference on Information Processing and Management of Uncertainty in Knowledge-Based Systems (IPMU 2002) (2002)
13. Lohweg, V., Diederichs, C., Müller, D.: Algorithms for hardware-based pattern recognition. EURASIP J. Appl. Sig. Process. **2004**(12), 1912–1920 (2004)
14. Lohweg, V., Voth, K., Glock, S.: A possibilistic framework for sensor fusion with monitoring of sensor reliability. In: Sensor Fusion, pp. 191–226. InTech (2011)
15. Mönks, U., Lohweg, V.: Machine conditioning by importance controlled information fusion. In: Proceedings of the 18th IEEE International Conference on Emerging Technologies and Factory Automation (ETFA 2013), pp. 1–8 (2013)
16. Mönks, U., Priesterjahn, S., Lohweg, V.: Automated fusion attribute generation for condition monitoring. In: Proceedings of the 23rd Workshop Computational Intelligence, vol. 46, pp. 339–353. KIT Scientific Publishing (2013)
17. Mönks, U., Trsek, H., Dürkop, L., Geneiß, V., Lohweg, V.: Assisting the design of sensor and information fusion systems. In: Proceedings of the 2nd International Conference on System-integrated Intelligence (SysInt 2014) (2014)
18. Mönks, U., Voth, K., Lohweg, V.: An extended perspective on evidential aggregation rules in machine condition monitoring. In: Proceedings of the 3rd International Workshop on Cognitive Information Processing (CIP 2012), pp. 1–6. IEEE (2012)
19. Niggemann, O., Stein, B., Vodencarevic, A., Maier, A., Kleine Büning, H.: Learning behavior models for hybrid timed systems. In: Proceedings of the 26th International Conference on Artificial Intelligence (AAAI 2012), pp. 1083–1090. AAAI (2012)
20. Osswald, C., Martin, A.: Understanding the large family of Dempster-Shafer theory's fusion operators - a decision-based measure. In: Proceedings of the 9th International Conference on Information Fusion, pp. 1–7 (2006)
21. Shafer, G.: A Mathematical Theory of Evidence. Princeton University Press, New Jersey (1976)
22. Verwer, S., Eyraud, R., Higuera, C.: Pautomac: A probabilistic automata and hidden markov models learning competition. Mach. Learn. **96**(1–2), 129–154 (2014)
23. Yadron, D.: Symantec develops new attack on cyberhacking: declaring antivirus software dead, firm turns to minimizing damage from breaches. Wall Street J., May 2014. published online at http://www.wsj.com/news/articles/SB10001424052702303417104579542140235850578
24. Zadeh, L.A.: Fuzzy sets. Inf. Control **8**(3), 338–353 (1965)

Text and Document Mining

Making Topic Words Distribution More Accurate and Ranking Topic Significance According to the Jensen-Shannon Divergence from Background Topic

Iwao Fujino[✉] and Yuko Hoshino

School of Information and Telecommunication Engineering,
Tokai University, Tokyo, Japan
{fujino,hoshino}@tokai.ac.jp

Abstract. This paper presents a useful approach for making topic words distribution more accurate and ranking topic significance according to the Jensen-Shannon divergence from background topic as a post-procedure of LDA method. In this paper, at first we defined the term score parameter to represent topics that will suppress the correlation between different topics and make the word distribution more accurate. Then according to the correlation between different topics, we described a concrete method for determining the proper setting of the number of topics. After that we proposed a method for ranking topic significance in the order of the Jensen-Shannon divergence from background topic. As a confirmation of our proposed methods, we conducted several experiments to processing English Twitter streaming data. The results of these experiments validate that our methods work efficiently as expected.

Keywords: Topic model · LDA (Latent Dirichlet Allocation) · Perplexity · Correlation coefficient · Jensen-Shannon divergence · Twitter streaming data

1 Introduction

1.1 Motivation

In recent year, with the tremendous advancement of internet technology and the wide prevalence of social networking service (SNS) and microblogging service, like Facebook and Twitter, people can report their daily activities and post their opinions to friends or public community personally anywhere and anytime. This circumstance brought a great opportunity of utilizing the massive information resource for modern society, but actually it caused a severe information environment for each individual user meanwhile. Because the amount of these information are too massive for users to access directly by themselves, it becomes necessary to ask computer intelligence system for help to classify or summarize the raw information to a sorted out form which may easy to access and easy to understand. Probabilistic topic modeling is an emerging approach that have been applied to explore and predict the underlying structure from massive data, especially from text documents. In topic model, it is assumed that

© Springer International Publishing Switzerland 2015
P. Perner (Ed.): ICDM 2015, LNAI 9165, pp. 185–200, 2015.
DOI: 10.1007/978-3-319-20910-4_14

documents can be considered as outward appearances of topics, which are a small set of latent variables corresponding to the underlying themes, causes or some kind of influences or forces behind the observed documents. The most useful implementation of topic model is LDA (Latent Dirichlet Allocation) method [1], which is an unsupervised machine learning technique to identify latent topic information from a massive document collection. In this method, sentences are decomposed into a set of words with no regard to the order of words and each document is represented in a list of words, which is prepared as input data of LDA processing. The output data of this processing are provided by a set of word probability distribution, which shows the appearance probability with respect to each topic for all words registered in the dictionary of the document collection.

Although LDA method is effective for extracting topics from a massive text document collection, there are still some important problems expected to be solved. The first problem is that does the LDA processing performed accurately or how to make more accurate LDA results for a given document collection. The setting of the number of topics is extremely critical with the accuracy and directly effects the interpretability of extracted topics. Furthermore the correlation between topics may contribute to insufficient performance of LDA processing results. The second problem is how to measure the significance of extracted topics. When utilizing LDA method to extract topics actually, some of the results reflect the topics behind the text data directly, but some of them maybe meaningless anyway. So it is expected to define an efficient significance measure, when ranking topics according to this measure, the higher ranking topics become more meaningful than lower ranking topics.

1.2 Our Study

In order to deal with these problems mentioned above, we contrived a useful approach for making the results of LDA method more accurate and more interpretable. Our study based on two basic assumptions. Because we think that a topic should be a cluster of similar information obviously distinguished from the other topics, as the first basic assumption, we think that there should be no correlation between any two different topics under ideal condition, i.e. the correlation coefficient between any two extracted topics should be zero. Furthermore we think that an ordinary distribution of words gathered from all documents can be considered to a virtual topic, which is called background topic. Because the background topic contains no special information at all, so that it should be the least significant topic of all topics. Therefore as the second assumption, we think that the significance of a topic should be defined as a measure that indicates the deviation from the background topic. As a measurement of deviation, we employ the Jensen-Shannon divergence between extracted topic and the background topic. In this paper, starting from these two basic assumptions, at first we defined term score parameter to represent topics that will suppress the appearance of common words and make the word distribution more accurate. Then we proposed a concrete method to determine the proper setting of number of topics when using LDA method to extract topics from documents collection. After that we proposed a method for ranking topics in order of a significance criterion according to the Jensen-Shannon divergence from background topic. As a confirmation of our proposed methods, we conducted several experiments to processing English Twitter streaming data. The results of these experiments validate that our methods work efficiently as expected.

1.3   Paper Outline

The remainder of this paper is organized as follows. In Sect. 2, we will provide some related work about processing the results of LDA method and ranking topics. In Sect. 3, as a preliminary, we will give a brief description about topic model and its LDA implementation. In Sect. 4, we will describe the basic concept and methodology of our study. We will provide a definition of term score for representing topics and show the property of improving accuracy, which is evaluated by perplexity. Then we will describe a concrete method for determining the proper setting of the number of topics. Furthermore we will provide the definition of the significance measure for ranking topics with the Jensen-Shannon divergence from the background topic. In Sect. 5, we will apply our proposed method to English Twitter streaming data and give the experiment results of ranking topics the significance measure. Finally in Sect. 6, we will conclude the achievements of our work and describe the some directions for our future research.

2   Related Work

Many previous works focus on post processing of LDA model and ranking topics have been reported. One of them is given by L. AlSumait et al. [2]. In their paper, they presented an automated unsupervised analysis of LDA models to identify and distinguish junk topics from legitimate ones and to rank the topic significance as well. Their basic idea consists of measuring the distance between topic distribution extracted by LDA method and a "junk topic distribution". They introduced three definitions of junk topic distribution in particular and they utilized a variety of metrics to compute the distance. Based on the results of distance calculation, they showed an expressive figure of topic significance which is implemented using a 4-phase weighted combination approach. They also tried some experiments to apply their ideas on synthetic and benchmark datasets. The results of these experiments showed the effectiveness of the proposed approach in ranking the significance of topics.

Another report is given by L. Wang et al. [3]. By combining the modified LDA_col (Latent Dirichlet Allocation_Collocation) model and re-ranking topics, they proposed an efficient method to find the topics which is easy to understanding for users. As the LDA_col can take word order into consideration, so that it becomes easy to discover more meaningful phases and more understandable topics. In this paper, they proposed a modified the LDA_col model to enforce consistency. In addition, they proposed two topic significance re-ranking methods, one is called topic coverage and the other is called topic similarity. As the result of their experiment, they assured that the topic re-ranking method based on topic coverage performs better than topic similarity method.

3 Topic Model and LDA Implementation

In topic model, document is described as bag-of-words without paying any attention to the order of word. Based on this assumption, PLSA (Probabilistic latent Semantic Analysis) has been proposed by T. Hofmann [4], and LDA has been proposed by D.M. Blei et al. [1]. As a preliminary, in the following, we will give a simple description about LDA implementation. Basically, for a given document set $W = \{w_1, w_2, \ldots, w_M\}$ where each document is given by $w_d = \{w_{d1}, w_{d2}, \ldots, w_{dN_d}\}(d = 1, 2, \ldots, M)$, its marginal distribution is can be expressed as follows if assuming that each document is generated independently.

$$p(W) = \prod_{d=1}^{M} p(w_d) \tag{1}$$

By introducing a topic proportion matrix as follows,

$$\theta = \begin{pmatrix} \theta_1 \\ \theta_2 \\ \vdots \\ \theta_M \end{pmatrix} = \begin{pmatrix} \theta_{11} & \theta_{12} & \cdots & \theta_{1K} \\ \theta_{21} & \theta_{22} & \cdots & \theta_{2K} \\ \vdots & \vdots & \ddots & \vdots \\ \theta_{M1} & \theta_{M2} & \cdots & \theta_{MK} \end{pmatrix} \tag{2}$$

the marginal distribution for each document w_d can be expressed as follows.

$$p(w_d) = \int p(\theta_d)p(w_d|\theta_d)d\theta_d \tag{3}$$

Because θ_d is chosen according to Dirichlet distribution, the first part in the integral of (3) can be expressed as follows, with a hyper parameter vector $\alpha = \{\alpha_1, \alpha_2, \ldots \alpha_K\}$.

$$p(\theta_d) = p_{Dir}(\theta_d; \alpha) = \frac{\Gamma(\sum_{k=1}^{K} \alpha_k)}{\prod_{k=1}^{K} \Gamma(\alpha_k)} \theta_{d1}^{\alpha_1 - 1} \cdots \theta_{dK}^{\alpha_K - 1} \tag{4}$$

The second part in the integral of (3) is given by

$$p(w_d|\theta_d) = p(w_d|\theta_d; \beta) = \prod_{n=1}^{N_d} \sum_{z_n} p(z_{dn}|\theta_d)p(w_{dn}|z_{dn}; \beta) \tag{5}$$

where $p(z_n|\theta_d)$ is distribution of topic variable and $p(w_n|z_n; \beta)$ is distribution of word variable with a hyper parameter matrix β as shown in (6), which is a word probability matrix for all topics and N is the number of unique word in the given document collection.

$$\beta = \begin{pmatrix} \beta_{11} & \beta_{12} & \cdots & \beta_{1K} \\ \beta_{21} & \beta_{22} & \cdots & \beta_{2K} \\ \vdots & \vdots & \ddots & \vdots \\ \beta_{N1} & \beta_{N2} & \cdots & \beta_{NK} \end{pmatrix} \tag{6}$$

Substituting (4), (5) into (3) and then substituting (3) into (1), finally we can obtain the marginal distribution for the document set W as follows,

$$p(W;\alpha,\beta) = \prod_{d=1}^{M} \int p_{Dir}(\theta_d;\alpha) \left(\prod_{n=1}^{N_d} \sum_{z_n} p(z_{dn}|\theta_d)p(w_{dn}|z_{dn},\beta) \right) d\theta_d \tag{7}$$

According to this mathematics expression, the LDA implementation can be depicted by a graphical representation as shown in Fig. 1.

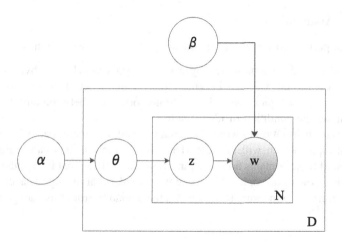

Fig. 1. Graphical model representation of LDA model

Then based on this graphical representation, document set can be generated by the following processing procedures.

(1) For each document $d = 1,2,\dots,M$:
   (a) Draw word distribution,
$$\theta_d \sim Dir(\alpha)$$

   (b) For each word $n = 1,2,\dots,N_d$:
       (i) Draw topic,
$$z_{dn} \sim Mult(\theta_d)$$
       (ii) Draw word,
$$w_{dn} \sim Mult(z_{dn})$$

where M denotes the number of documents and θ_d denotes the topic proportion in document d. Also Dir(.) denotes Dirichlet prior distribution and Mult(.) denotes multinomial distribution. After the inference of LDA, the results are represented as a distribution over words in which top probability words forms a semantically coherent concept and each document can be represented as a distribution over the discovered topics.

4 Basic Concept and Methodology

In this section, we will describe the basic concept and methodology of our study. We will describe the two basic assumptions of our study at first. We will provide a definition of term score for representing topics and derive the property of improving accuracy with respect to perplexity. After that we will explain the concrete methods for making LDA results more accurate and ranking topics in order of significance which is evaluated by the Jensen-Shannon divergence from the background topic.

4.1 Basic Assumptions

Our study is performed based on the following two basic assumptions.

1. The topic should be represented by a set of words which are obviously distinct words as contrasted with that in other topics. When a set of documents is categorized into a set of topics, there should be no correlation between any two different topics under the condition of ideal result.
2. An ordinary distribution of words gathered from all documents can be considered as a virtual topic, which will be called background topic in this paper. Because the background topic contains no special information at all, so that it should be the least significant topic of all topics. Therefore we consider that the significance of a topic can be defined as a measure that indicates the deviation from this background topic.

4.2 Term Score Representation for Topics

After extracting topics by LDA method, the results are provided by a word probability over all words in the dictionary for each topic. Although these high-ranking words indicate these words often appear in the topic, they are not typical words of the topic in the meaning of comparing with other topics because some of these words may also appear in other topics, so that relatively strong correlation may remain among topics. By following the standpoint of TFIDF weight parameter in information retrieval [5], we propose to use term score parameter to provide a topical words that shows specialty in comparison with other topics. Instead of the original word probability vector β_k,

$$\beta_k = \beta(t_k) = \begin{pmatrix} \beta_{1k} \\ \beta_{2k} \\ \beta_{3k} \\ \vdots \\ \beta_{Nk} \end{pmatrix} (k = 1, 2, \ldots, K) \tag{8}$$

we intend to represent topic t_k by means of term score vector $\boldsymbol{u}_k$ given as follows,

$$\boldsymbol{u}_k = \boldsymbol{u}(t_k) = \begin{pmatrix} u_{1k} \\ u_{2k} \\ u_{3k} \\ \vdots \\ u_{Nk} \end{pmatrix} (k = 1, 2, \ldots, K) \qquad (9)$$

where the element $u_{nk}(n = 1, 2, \ldots, N)$ denotes the term score for word n in topic k, which is calculated as follows,

$$u_{nk} = \beta_{nk}\log\left(\frac{1}{g_n}\right) = -\beta_{nk}\log g_n$$
$$(k = 1, 2, \ldots, K, n = 1, 2, \ldots, N) \qquad (10)$$

where g_n denotes the global term frequency for word n. If g_n is bigger, it means the word often appears in most documents of the document collection, this suppose that the word may be a common word for all topics. By multiplying $\log\left(\frac{1}{g_n}\right)$ to word probability β_{nk}, we can have a relatively smaller value of term score for each common word, therefore the common word will be removed from high-ranking of term score, and this results a uncorrelated representation for each topic.

4.3   Perplexity When Using Term Score Representation for Topics

Perplexity is a measure which indicates the prediction accuracy of words generalization performance for probability model. For a given document set $\boldsymbol{D}$ and each document $\boldsymbol{d} \in \boldsymbol{D}$, the mathematics expression of perplexity is given as follows.

$$perplexity(\boldsymbol{D}) = \exp\left(-\frac{\sum_{d\in D}\log p(\boldsymbol{d})}{\sum_{d\in D}N_d}\right) \qquad (11)$$

$$\log p(\boldsymbol{d}) = \sum_{w\in W}n_{dw}\log p(w|\boldsymbol{d}) \qquad (12)$$

$$p(w|\boldsymbol{d}) = \sum_{k\in K}p(w|k)p(k|\boldsymbol{d}) \qquad (13)$$

where N_d denotes the word amount in each document and n_{dw} denotes the count of word w appeared in document d. Perplexity is expressed by average of logarithm likelihood by probabilistic distribution $p(w|k)$. When using LDA method to extract K topics from test document set, we can express $p(w|\boldsymbol{d})$ as follows,

$$p(w|\boldsymbol{d}) = \sum_{k \in K} p(w|k)\, p(k|\boldsymbol{d}) = \sum_{k \in K} \beta_{wk}\, \theta_{kd} \tag{14}$$

where β and θ are denotation in the output results of LDA method. If we use term score for topic, we can substitute β_{wk} with u_{wk} which is given in (10), so that

$$p'(w|\boldsymbol{d}) = \sum_{k \in K} u_{wk}\theta_{kd} = \sum_{k \in K} (-\log g_w)\beta_{wk}\, \theta_{kd} = -\log g_w\, p(w|\boldsymbol{d}) \tag{15}$$

Then we can rearrange the expression of $\log p(\boldsymbol{d})$ as follows.

$$
\begin{aligned}
\log p'(\boldsymbol{d}) &= \sum_{w \in W} n_{dw}\log(-\log g_w\, p(w|\boldsymbol{d})) \\
&= \sum_{w \in W} n_{dw}\log(-\log g_w) + \sum_{w \in W} n_{dw}\log p(w|\boldsymbol{d}) \\
&= \sum_{w \in W} n_{dw}\log(-\log g_w) + \log p(\boldsymbol{d})
\end{aligned}
\tag{16}
$$

From the definition of perplexity as shown in (11), we can obtain the expression of perplexity for term score as follows,

$$
\begin{aligned}
perplexity(\boldsymbol{D}_{term\ score}) &= \left\{ \exp\left(\sum_{d \in D} \log p'(\boldsymbol{d}) \right) \right\}^{-\frac{1}{N}} \\
&= \left\{ \exp\left(\sum_{d \in D}\sum_{w \in W} n_{dw}\log(-\log g_w) + \sum_{d \in D}\log p(\boldsymbol{d}) \right) \right\}^{-\frac{1}{N}} \\
&= perplexity(\boldsymbol{D}_{word\ prob.}) \left\{ \exp\left(\sum_{d \in D}\sum_{w \in W} n_{dw}\log(-\log g_w) \right) \right\}^{-\frac{1}{N}}
\end{aligned}
\tag{17}
$$

where we denote $N = \sum_{d \in D} N_d$. Subsequently we can obtain the ratio of perplexity for term score and perplexity for word probability as follows,

$$
\begin{aligned}
\frac{perplexity(\boldsymbol{D}_{term\ score})}{perplexity(\boldsymbol{D}_{word\ prob.})} &= \left\{ \exp\left(\sum_{d \in D}\sum_{w \in W} n_{dw}\log(-\log g_w) \right) \right\}^{-\frac{1}{N}} \\
&= \left\{ \prod_{d \in D}\prod_{w \in W} (-\log g_w)^{n_{dw}} \right\}^{-\frac{1}{N}} \\
&= \exp\left(-\frac{1}{N}\left\{ \sum_{d \in D}\log \prod_{w \in W} (-\log g_w)^{n_{dw}} \right\} \right)
\end{aligned}
\tag{18}
$$

Then if assuming $log\left(\prod_{w \in W} (-\log g_w)^{n_{dw}} \right) > 0$, we can obtain

$$\frac{perplexity(\boldsymbol{D}_{term\ score})}{perplexity(\boldsymbol{D}_{word\ prob.})} < 1 \tag{19}$$

At last for a sufficient condition, this is equivalent to $\log g_w < 1$ or $g_w < e$ for $w \in W$. This means we can achieve an improvement of perplexity if we can satisfy this sufficient condition.

4.4   Determining the Number of Topics

According to the first assumption, we sorted out our consideration for determining the number of topics as follows. When we assigned a proper value to the number of topics, the correlation between any two different topics should be quite small. When assigned a value smaller than that of the latent topic in document collection, a latent topic may be divided into two different topics in the result of LDA method, this may causes the rising of correlation. On the other side, when assigned a value bigger than that of the latent topic in document collection, under the restriction of finite words of given document collection, same words may appeared in the result of different topics, this may causes the rising of correlation as well. Eventually we understand that the graph of the correlation between different topics for a series setting of the number of topics should form a concave curve and the proper value of the number of topics should the value corresponding to the minimum correlation. Subsequently in order to provide a scalar parameter to evaluate the correlation for each setting of the number of topics, we intent to utilizing the maximum value of the correlation coefficient between any two different topics for considering the worst situation.

4.5   Ranking Topics in Order of Significance

Before discuss ranking topics in order of significance, we will give a definition on the significance of topic. If a topic is steadily talked for a very long period, it may not attract users' concern at all, in other words, users may lose interest in this topic. From this standpoint, if we represent the ordinary word distribution of all documents as a virtual topic, called background topic, we think the background topic contains no special information to all users, so it should be the least significant topic of all topics. Therefore the significance of a topic should be defined as a measure that indicates the deviation from the background topic. In this paper, we use v to denote the word probability vector of background topic as follows.

$$v = \begin{pmatrix} v_1 \\ v_2 \\ v_3 \\ \vdots \\ v_N \end{pmatrix} \tag{20}$$

where $v_n (n = 1, 2, \ldots, N)$ indicates the word probability under ordinary situation. According to the definition of Jensen-Shannon divergence, the deviance between u_k $(k = 1, 2, \ldots, K)$ and v can be expressed as follows,

$$jsd(u_k, v) = \sum_{n=1}^{N} \left(u_{nk} log \frac{u_{nk}}{q_{nk}} + v_n log \frac{v_n}{q_{nk}} \right) (k = 1, 2, \ldots, K) \tag{21}$$

where $q_{nk} = \frac{1}{2}(u_{nk} + v_n)$. Because Jensen-Shannon divergence shows the deviance between two probability distributions, the bigger this measure is, the more deviation there are between topic k and the background topic v are.

5 Application to Twitter Streaming Data

In order to perform a confirmation experiment, we apply our proposed approach to Twitter streaming data. In this section, we will describe the experiment conditions and report the results of our experiments.

5.1 Experiment Conditions

In our study, as the first step, we gathered English tweets provided by the sample level of Twitter streaming API and accumulated these data to CouchDB separately according to the crawling date. The summary of the document collection we used for confirmation experiment is shown as follows (Table 1).

Table 1. Summary of document collection

Database name	Entwitter20141118	Entwitter20141119
Crawling date	2014/11/18	2014/11/19
Size in CouchDB	9.3 GB	9.2 GB
Number of documents	1,421,712	1,397,883
Number of effective authors	13,223	13,496
Number of effective words	6,780	6,736

In order to conduct the experiments more precisely, we performed a four steps preprocessing shown as follows:

1. To prepare a relatively bigger data chunk for using LDA method, we cluster gathered tweets according to user's account name.
2. To make the LDA model work more effectively, we only choose the tweets which length is more than 28 words as effective tweets.
3. To prepare the data for LDA method, we count words appeared in each document and save the result to text file according to Blei's format.
4. To prepare the global term frequency, we count words appeared in document collection crawled for a relatively long period and save the result to text file.

5.2   Experiment Results of Calculating Perplexity

The first experiment is about to confirm the perplexity of proposed term score representation for topics. By applying LDA program released by D. Blei[1] to the document collection prepared in Subsect. 5.1, we conducted LDA processing experiment varying the setting of the number of topics from 10 to 80 for document collection entwitter20141118 and entwitter20141119. From the result of each setting we obtained the word probability distribution for each topic and the topic proportion for each document. Then we calculated the term score according to (9) and (10). Besides using these results we calculated the perplexity according to (11)–(13) for each setting. At last we achieved a curve graph of perplexity vs. the number of topics, which is shown in Fig. 2. This curve graph gives a comparison between the word probability case and the term score case for different document collections. From this graph we can understand two things. The first thing is that the perplexity value in the term score case is quite small than that in word probability case throughout the range of the number of topics for either document collection. The second thing is that the results of perplexity of the two document collections show same tendency, with the ratio of perplexity of term score and that of word probability shown in (18). Therefore we assert that the term score representation of topics is more accurate than word distribution representation of topics.

5.3   Experiment Results of Correlation Coefficient Between Topics

By adapt LDA model to the document collection entwitter20141118, we can obtain a set of topics, which is provided by a word probability list for each topic and we can put the word probability result in vector form according to (8). Then employing this word probability vector, we can calculate the matrix of correlation coefficient between any two topics according to the following mathematics expression in this case:

---

[1] http://www.cs.princeton.edu/ ~ blei/lda-c/index.html.

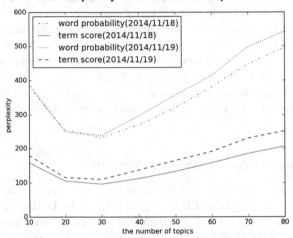

Fig. 2. Perplexity carve of the case of word probability and term score

$$R_\beta = \left(r(\beta_i, \beta_j)\right) = \left(\frac{\sum_{n=1}^{N} \left(\beta_{ni} - \bar{\beta}_i\right)\left(\beta_{nj} - \bar{\beta}_j\right)}{\sqrt{\sum_{n=1}^{N} \left(\beta_{ni} - \bar{\beta}_i\right)^2}\sqrt{\sum_{n=1}^{N} \left(\beta_{nj} - \bar{\beta}_j\right)^2}}\right) \tag{22}$$

$$(i = 1, 2, \ldots, K, j = 1, 2, \ldots, K)$$

The calculation result is depicted in 3D graph shown in Fig. 3. The straight line in the upper part of Fig. 3 shows correlation coefficient of any topics themselves, which value is equal to 1 exactly, i.e. $r(\beta_i, \beta_i) = 1 (i = 1, 2, \ldots, K)$. The lower part of Fig. 3 shows correlation coefficient of between any two different topics, which have value of less than 1, i.e. $r(\beta_i, \beta_j) < 1 (i \neq j, i = 1, 2, \ldots, K, j = 1, 2, \ldots, K)$. We can see that there are many points protruding from lower plane, where the value of correlation coefficient are bigger comparatively and the maximum value of $r(\beta_i, \beta_j) (i \neq j)$ is 0.92.

In addition, instead of employing the word probability parameter vector, we calculated correlation coefficient matrix employing the term score parameter vector according to (9) and (10). The mathematics expression of correlation coefficient matrix is given as follows in this case:

$$R_u = \left(r(u_i, u_j)\right) = \left(\frac{\sum_{n=1}^{N} \left(u_{ni} - \bar{u}_i\right)\left(u_{nj} - \bar{u}_j\right)}{\sqrt{\sum_{n=1}^{N} \left(u_{ni} - \bar{u}_i\right)^2}\sqrt{\sum_{n=1}^{N} \left(u_{nj} - \bar{u}_j\right)^2}}\right) \tag{23}$$

$$(i = 1, 2, \ldots, K, j = 1, 2, \ldots, K)$$

The calculation result is depicted in 3D graph shown in Fig. 4.

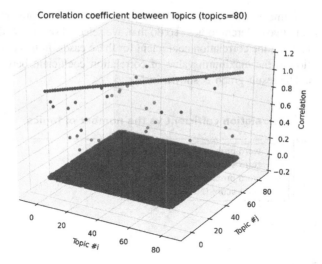

Fig. 3. Correlation coefficient between any two topics (Employing word probability for representing topics)

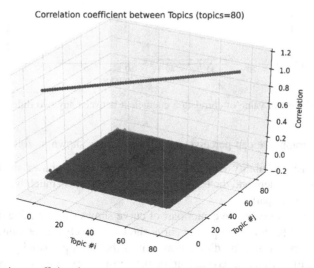

Fig. 4. Correlation coefficient between any two topics (Employing term score for representing topics)

The straight line in the upper part of Fig. 4 shows correlation coefficient of any topics themselves, which value is equal to 1 exactly. The lower part of Fig. 4 shows correlation coefficient of between any two different topics, which have value of less than 1. We can see that there are several points protruding from the lower plane the maximum value is 0.17 this time. Comparing Fig. 4 with Fig. 3, We can perceive that the value of correlation coefficient between any two different topics becomes much smaller in this graph. As a result, it can be confirmed that results of topics are uncorrelated significantly when employing term score for representing topics.

The next experiment is about to check the performance of suppressing the correlation between any two different topics. To do that, we selected several settings of topic number and calculated the correlation coefficient for these cases. In this experiment, we only pay attention to the maximum value of correlation coefficient between any two different topics. The result graph is shown in Fig. 5.

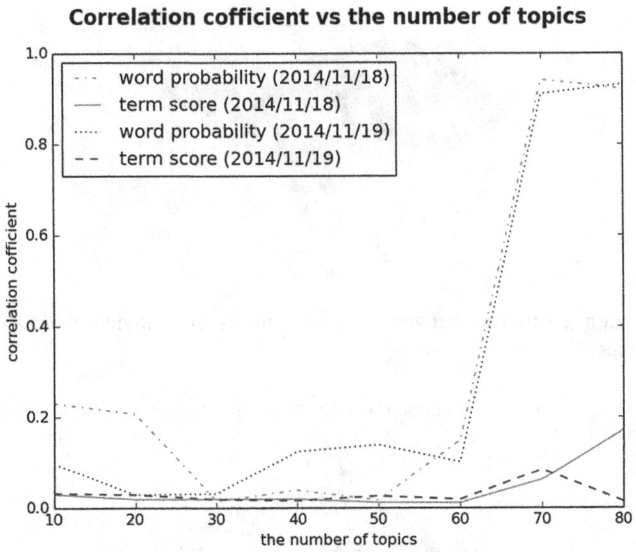

Correlation cofficient vs the number of topics

Fig. 5. Maximum value of correlation coefficient between any two different topics

From this graph, we can perceive several particulars shown as follows:

1. For the word probability case, the center part of curve shows smaller value of correlation coefficient and the both side part of curve shows much bigger value than that of the center part.
2. For the term score case, the center part of curve shows smaller value of correlation coefficient and the both side part of curve shows relatively bigger value than that of the center part, but not too different as that of the word probability case.
3. All values of correlation coefficient in the term score case are smaller than its corresponding value in the word probability case, especially at the both side of the curve.
4. From the standpoint of correlation, the proper setting of the number of topics should be determined in the range of 30 to 50, and 40 may appropriate in this case.

5.4 Experiment Results of Ranking Topics in Order of Significance

As a confirmation experiment, we attempted to rank the topic results from Twitter stream data, using the method proposed in Subsect. 4.5. The long period global term

Table 2. Result of ranking topics by Jensen-Shannon divergence from background topic

Rank	Topic no.	Jsd	Rank	Topic no.	Jsd
1	5	17.94818	21	7	17.431413
2	19	17.793257	22	23	17.429183
3	16	17.749459	23	0	17.415109
4	35	17.739493	24	2	17.404172
5	33	17.729825	25	24	17.393955
6	12	17.704564	26	20	17.389982
7	22	17.688723	27	28	17.365673
8	15	17.662412	28	39	17.364891
9	3	17.619348	29	29	17.362601
10	25	17.603029	30	11	17.352194
11	9	17.574173	31	1	17.346253
12	34	17.571173	32	4	17.344147
13	37	17.553141	33	30	17.338528
14	14	17.528044	34	13	17.326589
15	21	17.506123	35	31	17.298277
16	10	17.464791	36	36	17.296353
17	26	17.45997	37	27	17.261744
18	8	17.455114	38	6	17.209833
19	18	17.454345	39	32	17.16954
20	17	17.44525	40	38	17.11882

Table 3. Top 15 typical words result of background topic and the top 6 topic ranked by proposed approach

Background topic	Rank 1	Rank 2	Rank 3	Rank 4	Rank 5	Rank 6
	Topic #5	Topic #19	Topic #16	Topic #35	Topic #33	Topic #12
Day	Fun	Car	Ass	Week	Bit	God
Time	Brazil	House	Dick	Grade	Talk	Game
Life	Day	Snow	You	Birthday	Cat	Hug
People	Girlfriend	Thought	Gym	Teacher	Brother	Child
Girl	Butt	Soul	Pussy	Book	Player	Freshman
Love	Happy	Figure	Everybody	Weekend	Deal	Self
Year	And	Mess	Start	Chase	Lil	Password
Thing	Ice	Street	Follower	Concert	Truth	Waiter
Man	Never	Drift	None	Step	Call	Beer
Way	Contract	Winter	Tattoo	Monday	Window	Fault
Guy	Should	Degree	Run	Gas	Luck	Board
World	Way	Decision	Pack	Waste	Floor	Bug
Lot	Roll	Cold	Reality	Lies	Son	Dah
Friend	Size	Sign	Over	Worth	Beat	Flight
Shit	Dress	Blessing	Come	December	Memory	Passion

frequency is gathered from November 16, 2014 to December 12, 2014. The target document collection is gathered in November 18, 2014. For the case of employing Jensen-Shannon divergence to measure the deviation from ordinary situation, the calculation results according to (20) and (21) is shown in Tables 2 and 3. From these results we can perceive that the high-ranking topics are more interpretable and easy understanding.

6 Conclusion and Future Work

In this paper, we presented a useful approach for making topic words distribution more accurate and ranking topic significance according to the Jensen-Shannon divergence from background topic as a post processing procedure of LDA method. Our study based on two basic assumptions. The first assumption that there should be no correlation between any two different topics under ideal condition. The second assumption is that the significance of a topic should be defined as a deviation measure from the background topic. Based on these assumptions, we defined the term score parameter to represent topics that will suppress the correlation between different topics and make the word distribution more accurate. Also according to the correlation between different topics, we described a concrete method for determining the proper setting of the number of topics. Moreover we proposed a method for ranking topics in order of a significant criterion according to the Jensen-Shannon divergence from background topic. As a confirmation experiment, we applied the proposed approach to English Twitter streaming data, and we can find that most of the high-ranking topics are easy to understand. For future work, by adapt the proposed approach of this paper to DTM (Dynamic Topic Model) processing results, we will try to discover the variations of topic's ranking according to the passage of time.

References

1. Blei, D.M., Ng, A.Y., Jordan, M.I.: Latent Dirichlet allocation. J. Mach. Learn. Res. **3**, 993–1022 (2003)
2. AlSumait, L., Barbará, D., Gentle, J., Domeniconi, C.: Topic significance ranking of LDA generative Models. In: Buntine, W., Grobelnik, M., Mladenić, D., Shawe-Taylor, J. (eds.) ECML PKDD 2009, Part I. LNCS, vol. 5781, pp. 67–82. Springer, Heidelberg (2009)
3. Wang, L., Wei, B., Yuan, J.: Topic discovery based on LDA_col model and topic sinificance re-ranking. J. Comput. **6**(8), 1639–1647 (2011)
4. Hofmann, T.: Probabilistic latent sematic analysis. In: UAI, pp. 289–296 (1999)
5. Baeza-Yates, R., Ribeiro-Neto, B.: Modern Information Retrieval – the Concepts and Technology behind Search, 2nd edn. Pearson Education Limited, Harlow (2011)

Visualized Episode Mining with Feature Granularity Selection

Sonja Ansorge$^{(\boxtimes)}$ and Jana Schmidt

Gesundheitsforen Leipzig GmbH, Hainstr 16, 04109 Leipzig, Germany
{ansorge,schmidt}@gesundheitsforen.net

Abstract. There has been much effort in the last decade to provide of the shelf software for episode mining, but still it remains a challenge. This is especially true for sequences that contain hierarchical events, e.g. diagnosis codes. Then, the user must not only decide which features should be used, but also in which granularity. This is even complicated due to the fact that, visualizations of the results of episode mining are rare. Therefore, we first introduce an extension to LifeLines2 that is able to mine and visualize sequential data and second, propose a feature granularity selection method to handle hierarchical event sequences. The granularity selection of features applies a greedy bottom up search that also incorporates support and run time constraints. A significant run time improvement is achieved by a parallelization approach. Finally, we show the applicability on real world medical data sets.

Keywords: Episode rule mining · Hierarchical event histories · Medical data mining · LifeLines2 · dmt4sp

1 Introduction

Temporal data mining can be defined as the extraction of non-trivial information from large sequential data sets. There, sequential data are ordered with respect to some kind of index (e.g. time). This type of data mining differs from the traditional and well researched time series analysis by the size and nature of data sets. The objective of temporal data mining is usually the detection of interesting trends and patterns, whereas typical applications in time series analysis include forecast and control or model parameter calculations. Furthermore, the ordering in the data makes temporal pattern mining an especially challenging field. The patterns to be found are called episodes [1], an ordered set of events, which occur in rather short succession. Frequent episode and episode rule mining is the equivalent of frequent item-set and association rule mining with the additional dimension of time. In contrast to the well researched topic of association rule mining, approaches for episode mining are still rare [2]. However, the command line tool dmt4sp [3], which is a Data Mining Tool for Serial Pattern allows for mining of frequent serial episodes and serial episode rules, under a variety of syntactic (e.g. pattern size, fixed target event), time (e.g. gap

© Springer International Publishing Switzerland 2015
P. Perner (Ed.): ICDM 2015, LNAI 9165, pp. 201–215, 2015.
DOI: 10.1007/978-3-319-20910-4_15

and window sizes), support and confidence constraints. One application of temporal data mining are electronic health records (EHR). Event types recorded in this setting include diagnoses, medication prescriptions, laboratory findings or performed therapies. Although, visualization of medical data is already a widespread practice in the health care system, analysis of the relationship of the recorded events is not practiced frequently, yet. Only simple relationships between attributes, leading to rule discovery or regression models are state of the art. Additionally, existing tools often only serve one purpose: either visualization or data mining of event sequences. In contrast, we believe that an interactive visualization tool combined with a data mining tool will allow the user to exploit their perceptive and cognitive capabilities for additional exploration and analysis of the data. This may include better comprehension of data and pattern mining results, hypothesis generation and evaluation, as well as knowledge discovery. But not only the combination of these two domains is beneficial for the analysis of medical records. Usually, in the analysis of data including diagnosis codes the user can choose on which level of granularity they are used. This is due to the special structure of such information. They can be given in different granularities, from very rough to very detailed along with all benefits and merits. Much effort is spent to find an optimal level of granularity leading to good results. We also want to address this problem by introducing an automated feature selection procedure that is able to identify a good level of granularity. Thus, the user only gives constraints for this search and is returned a set of attribute levels giving in valuable results. Consequently, the goals of this work are

1. the combination of visualization and data mining tools into a framework for the interactive visual and computational analysis of timed categorical data and patterns,
2. the additional inclusion of a new feature selection method with focus on the hierarchical event structure
3. and a performance improvement through the means of parallelization.

This paper is organized as follows. Section 2 reviews related work. Section 3 gives an overview of episode mining and the problem for hierarchical data sets. Moreover, our proposed solution is introduced along with its inclusion in LifeLines2. The following Sect. 4 shows how this inclusion works on synthetic and real world data sets. The paper closes with a discussion.

2   Related Work

Visualization for Medical Data. Visualization of medical data is very important in clinical settings and covers areas as diverse as disease outbreaks [4] or drug side- and cross effects [5]. Moreover, due to the growing importance of EHRs, a number of different tools have been introduced for this type of temporal categorical data. Many of these tools focus on the data of a single patient, including LifeLines [6] or Web-Based Interactive Visualization System [7]. Their aim is generally to support physicians in obtaining a quick overview of usually lengthy patient histories. Moreover, there are a number of systems that focus on search and aggregation strategies of multiple EHRs, for example Similan [8], and VIS-ITORS [9]. However, as of our knowledge the introduction of data mining into such visualization tools has not been performed, yet. Therefore, our first aim is to combine a data visualization tool with a data mining software.

Feature Selection for Hierarchical Event Sequences. Feature selection specifies the process when a set of predefined features has to be chosen that covers all necessary information for data mining. Many approaches have been developed for this task and share the goal to select only attributes that contribute to the target variable [10]. However, in this paper, the feature selection task is different. As outlined in detail in Sect. 3.3 attributes are given on a very specific level and moreover, are organized in a tree-like structure. Depending on the data mining task different levels of granularity may be appropriate. Therefore, the feature selection method should automatically find an appropriate set of attributes out of the tree-like structured attributes and not automatically retrieve the lowest level. Especially in the domain of medical data mining such attribute types are present [11,12]. Nevertheless, to the best of our knowledge, no-one already addressed this issue - such attributes were always used in the best granularity. Thus, the second goal is to provide a first step towards the correct selection of granularity on sequences with hierarchical events.

3   Methods

This section briefly introduces LifeLines2 and the basics of episode rule mining (ERM). Subsequently, we present our approach on the problem setting of hierarchical feature selection.

3.1   LifeLines2

LifeLines2 is an interactive visualization tool from the University of Marylands Human-Computer Interaction Laboratory. It was created for the inspection of electronic health records, but in general, all sequence data can be loaded. Additionally LifeLines2 provides sophisticated means to visualize, manipulate, and summarize multiple records at once. Data can be explored with the align, filter and rank tools as well as condensed via temporal summaries. Moreover, the filter function includes a temporal pattern search algorithm [14] that facilitates filtering patient groups for the occurrence of user defined event sequences. However, no data mining algorithms are included in this tool, yet. For a detailed description of Lifelines2 please refer to its homepage[1], where also an excellent tutorial is given [13].

---

[1] http://www.cs.umd.edu/hcil/lifelines2/.

3.2   Episode Mining

In this work we follow the standard definitions of event sequence, episode, occurrences and support as introduced by Mannila et al. [1]. Let E be a set of *event types*, respectively the event type *alphabet*, e.g. $E = A, ..., Z$. An *event* is a pair (e, t), e.g. $(A, 4)$, where $e \in E$ is the event type and t is the *time-stamp* which denotes the occurrence time of the event. Time-stamps are usually given with integers, but the interval between two time steps can be associated with an arbitrary length of time (e.g. minutes, days or months). An *event sequence* or *event history* $s = \langle (e_1, t_1), (e_2, t_2), ..., (e_n, t_n) \rangle$ is an ordered sequence of events such that $\forall i \in \{1, ..., n\}, e_i \in E \wedge t_i \in \mathbb{N}$ and $\forall i \in \{1, ..., n-1\}, t_i \leq t_{i+1}$, e.g.: $\langle (D, 1), (A, 12), (B, 15), (C, 25), (E, 25), (A, 36), (B, 38), (C, 55), (D, 66), (E, 75) \rangle$. A categorical, temporal data set may consist of one (usually longer) event sequence or several independent event sequences. In the real world data used in this work one event sequence corresponds to one medical patient record and a data set equates to a patient population or patient group. A *serial episode* α, in the following referred to as episode, is a succession $\langle e_1, e_2, ..., e_n \rangle$ of event types, e.g.: $\langle A, B, C \rangle$. The elements of α have a strict order such that e_1 is followed by e_2, which is followed by e_3 and so forth. The *size* of α, i.e. $|\alpha|$, is equal to the number of elements contained in the episode (n). An episode α *occurs* in an event sequence s, if there is at least one ordered sequence of events $s' = \langle (e_1, t_1), (e_2, t_2), ..., (e_n, t_n) \rangle$, such that s' is a sub sequence of s, i.e. s' can be obtained by removing some (or no) elements from s. The *occurrence* $T = [t_1, t_2, ..., t_n]$ is equal to the time-stamps of the matching sub sequence s'. In the example sequence there are four occurrences of the episode $\langle A, B, C \rangle$: $T_1 = [12, 15, 25]$, $T_2 = [12, 15, 55]$, $T_3 = [12, 38, 55]$ and $T_4 = [36, 38, 55]$. A *minimal occurrence* of an episode α, i.e. an occurrence that does not contain any other occurrences of the same episode, is defined as follows: If $|\alpha| = 1$, then all occurrences of α are minimal. Otherwise, an occurrence $T = [t_1, t_2, ..., t_n]$ is minimal iff there is no other occurrence $T' = [t'_1, t'_2, ..., t'_n]$ such that $(t_1 < t'_1 \wedge t_n = t'_n)$ or $(t_1 = t'_1 \wedge \exists i \in \{2, .., .n\}, t'_i > t_i)$. In the example sequence there are two minimal occurrences of the episode $\langle A, B, C \rangle$: $T_1 = [12, 15, 25]$ and $T_2 = [36, 38, 55]$. The *support* of an episode or episode rule in a data set can either be defined by the number of minimal occurrences or by the number of event sequences which contain at least one minimal occurrence. In the following we will use the latter definition, which is sometimes also denoted as *sequence support*. A *serial episode rule*, in the following referred to as episode rule, is defined as the expression $\alpha \Rightarrow e_{n+1}$, e.g.: $\langle A, B, C \rangle \Rightarrow D$. This can be read as 'the occurrence of episode α (*antecedent*) implies that the *target event* (*consequent*) e_{n+1} will follow'. The (minimal) occurrence and support of an episode rule are equal to the (minimal) occurrence and support of an extended episode $\alpha' = \langle \alpha, e_{n+1} \rangle$. The *confidence* of a rule $\alpha \Rightarrow e_{n+1}$ is equal to the number of minimal occurrences of α' divided by the number of minimal occurrences of α. In the example the rule $\langle A, B, C \rangle \Rightarrow D$ has two occurrences, $T_1 = [12, 15, 25, 66]$, $T_2 = [36, 38, 55, 66]$, where only the second is a minimal occurrence. The rule body however, has two minimal occurrences. Thus the confidence of this rule is 0.5.

Dmt4sp. The Data Mining Tool for Serial Pattern [3] is a command line tool for the detection of frequent serial episodes, episode rules and quantitative episodes. In this work we focused on the former two options. Dmt4sp can either be applied to a single sequence of events or, as in our case, to a set of event sequences. Several events may occur at the same time as long as their type is different. Patterns may be extracted under a variety of constraints (support, confidence, time intervals or syntactic constraints), all of which may be combined leaving only patterns satisfying all constraints as a result. Note that the confidence value used by dmt4sp is based on minimal occurrences an therefore different from the confidence used in later sections.

3.3   Problem Setting: Feature Granularity Selection of Hierarchical Event Histories

After having introduced ERM, we present the problems with feature granularity selection that arise when a hierarchical data set is given. Then, a solution for this step is introduced.

Hierarchical Event Histories. In this paper, we present a method to automatically find a good set of hierarchical events. Hierarchical means that the input events can be used in different granularities, ranging from specific to general. Consider for example diagnosis codes (ICD-codes), which are used by physicians to code the patients diseases, where I63, for example, is cerebral infarction, while I63.1 is cerebral infarction due to an embolus. ICD codes consist of up to five levels of hierarchy, which leads the user to the problem to define on which level should be used. Figure 1 illustrates this problem setting. Events can be used on any granularity level, where it is not necessary that all sub-event are on the same level.

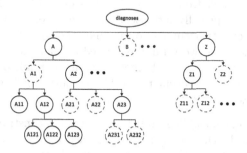

Fig. 1. Illustration of hierarchical events. Events are ordered in a tree like structure, ranging from general to specific. The red dashed circles indicate the set of attributes that lead to a good F-measure of the resulting episodes and thus should be selected by the algorithm

In medical data not only diagnoses are built up in this way, but medication codes (ATC-codes) as well. Of course there may be many other domains, where such hierarchical features occur. Let's consider a data set of events as defined in Sect. 3.2. However, the events that are recorded in this data set can be used at different granularity. The aim of the feature granularity selection process is to identify an optimal set of features A' from all features $A = \{a_1, \ldots, a_n\}$ on different granularities so that the retrieved episode rule set $E(A) = \{\epsilon_1, \ldots, \epsilon_k\}$ shows a high support but also a high confidence, which is captured by the F-measure $F_{E(A)}$ (cf. Eq. 2): $A' = \text{argmax}_A \, F_{E(A)}$ In the following, this process is described.

Feature Granularity Selection. In this section we show the introduction of a hierarchical feature selection process into LifeLines2. In typical episode mining applications events can either be included or not included into the analysis process. However, the hierarchical structure of events (e.g. ICD codes) allows for a third option: the inclusion of a higher, less detailed, hierarchy level instead of the full event code. Therefore a feature selection process must not only determine which events are informative, but also which level of the hierarchy is the best choice. To allow for a supervised feature selection the consequent of the mined episode rules has to be a predefined target event, splitting the available amount of patients into a positive class, where the event occurs, and a negative class, where the event does not occur. The user interface for the feature selection process was incorporated in new episode mining interface (see Fig. 6). All episode rule mining options like minimum support or target event are applied when given. Note that the feature selection cannot be performed without a target event and an additional record file that contains instances of the negative class.

The feature selection process generally starts from the lowest level of the hierarchy. There is, however, one important fact to be considered here: The support of an episode rule is always equal to or less than the lowest support of all contained events. Therefore event types that do not comply with the minimum support can be excluded from the mining process without loss of important information. This is especially interesting for the lower levels of the hierarchy since very specific events occur less often than their generalized counterparts. Accordingly, we check for every fully expanded event node, whether it satisfies the minimum support constraint. If the constraint is fulfilled the event type is included in the initial set of event types. If the constraint is not satisfied there are two possibilities (cf. Fig. 2 for an example):

1. The event node has no sibling node that fulfills the constraint. Then, instead of the current node, the parent node is checked for its support.
2. The event node has a sibling node that fulfills the constraint. Then, the current node is simply exempt from the initial set of event types.

Starting from the so determined initial event types, nodes are gradually collapsed in the ascending order of support values. That means that rare event types are collapsed before frequent ones, allowing for more common patterns involving these events. When an event node is collapsed, the parent event is included into the current set of selected features and all children are excluded. To make sure that the predefined target event is contained in the mining process the path to

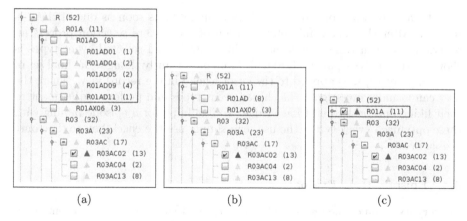

(a)   (b)   (c)

Fig. 2. Determination of the starting event set for feature selection at the example of ATC-code 'R' (Respiratory System). The minimum support used here is ten. As can be seen in a) all event types with the prefix 'R01AD' have support values lower than ten, thus the parent event 'R01AD' is tested for the minimum constraint support instead. Neither the parent event nor its sibling 'R01AX06' satisfy the constraint (b), so now their common ancestor 'R01A' is tested instead. This node finally complies with the constraint, thus it is included into the initial feature set (c). However, the event types 'R03AC04' and 'R03AC04' fall into the second testing category. They do not fulfill the support constraint, but they have a sibling event type node, 'R03AC02', that does, so they are both excluded from the initial feature set.

the target event always stays fully expanded. After every collapse, the current set of features is used for episode rule mining on the positive data set and the resulting rules are then evaluated (see Sect. 3.4). If the latest collapse improved the mined episode rules this collapse is kept. Otherwise the collapse is undone and a different node will be collapsed in the next iteration. A collapse that was undone once will never be performed again, but the collapse of more general ancestor event types is still possible. E.g.: assume that the collapse of event type 'ATC R03AC' was rejected and will not be performed again, since it did not bring any improvement. Then the ancestor event type 'ATC R03A' may still be collapsed. Both our bottom up approach and the rule, that nodes only stay collapsed when this improves the results, are based on the following assumption: It is a fact that the less specif an event type, the higher the support and sequence support. We assume that this leads to more valid event combinations for the mined episode rules, but also to considerably longer run times. To avoid overly long run times and still enable improvement through the feature selection, we thus tried to keep the event types as specific as possible and as general as necessary. The feature selection process ends automatically when all possible collapses have been performed. However, since we expect long run times for this process, we implemented three additional user defined termination criteria: The maximum number of iterations, the maximum run time for a single iteration and the maximum total run time. The user may set some or all of these criteria and the feature selection process will be terminated as soon as one of them is reached. After the successful determination of the feature selection process an overview over run time, iterations, quality criteria and selected event types is shown in the analysis panel (cf. Fig. 6). More importantly though, the final set of selected event types is applied to the event legend in the record view. There the user can visually investigate the chosen event types and perform adjustment to their liking and background knowledge. To finally perform episode mining with these optimized event types, the user can simply use the check-box to 'Use only visible events'.

3.4   Episode Rule Evaluation

In every iteration of the feature selection procedure a different set of event types is tested for their capacity to produce 'better' episode rules than the previous set of events. Only the dmt4sp specific minimum support and confidence constraints are used to sort out rules during the mining process. No rules are removed after the mining process, instead the produced rule set as a whole is evaluated. In the following section we explain how the predictive capacity and therefore quality of a set of episode rules can be measured.

In order to measure this quality, rules have to be learned first (see step 1 in Fig. 3). Since the target of the episode rules is already predefined only positive instances C^+, i.e. histories containing the target event, need to be included in the episode rule mining process. After the mining is finished we measure the predictive quality of each episode rule separately using the standard quality measure of *confidence*. The confidence measures how well an occurrence of the rule body predicts a following occurrence of the target event. The confidence is defined as the number of histories containing the rule (n_r) divided by the number of histories containing the body (n_b).

$$Confidence = \frac{n_r}{n_b} = \frac{n_r}{n_b^+ + n_b^-} = \frac{n_r}{n_r + n_b^-} \tag{1}$$

Note that this confidence value is different from the one used by the dmt4sp. Occurrences of the full rule can be calculated from C^+ alone, since per definition the target event does not occur in the set of negative instances C^-. Occurrences of the rule body can be split into occurrences from positive (n_b^+) and negative instances (n_b^-). Since the histories in C^+ are cut off after the first occurrence of the target event prior to the episode mining process, n_b^+ is equal to n_r. The only missing information to calculate the confidence is n_b^-. Thus every history in C^- needs to be searched for any matching episodes rule bodies (see step 2 in Fig. 3). For this step the built in LifeLines2 pattern search algorithm is used. After this step is completed, confidence values will be assigned to all rules. In order to assess the quality of the resulting episode rule set as a whole, the remaining histories of the mixed test set are used. Every history is assigned a *score* between zero and 100 according to the highest confidence of any matching episode rule body. A score of zero is assigned when no episode matches. In an

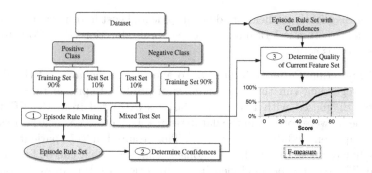

Fig. 3. Work-flow of the feature selection procedure. The full data set is first split into a negative and a positive class according to the occurrence of the chosen target event. Subsequently, two separate training sets and a mixed test set are formed. The following steps one to three are repeated during every iteration of the feature selection procedure. In the first step the positive training set is used to learn episode rules with the dmt4sp algorithm. During the next step the rule confidences are determined by scanning the negative training set. During the third and last step scores are assigned to test set and used together with the score threshold to predict the class of every instance. From this prediction the quality of the current event type set is calculated. If the quality improved, the current feature set is used as the basis of the next iteration.

ideal scenario the score translates directly into the probability that the target event of the episode rule will occur later in the history, i.e. a high score equals a high chance for the test instance to be positive. In order to calculate these scores a third searching step is necessary where every history in the test set is scanned for any matching episode rule bodies (see step 3 in Fig. 3). Subsequently, a user defined score threshold (default 80) is used to predict the class of every test instance. For this prediction we calculate the quality measures *precision* and *recall*. Finally, to compare two different event type sets and their corresponding episode rule sets during the feature selection procedure we use the *F-measure* [10]. Only when the F-measure of the current set is larger than currently best F-measure, this set will be used as the basis of the next iteration.

$$F\text{-}measure = 2 \cdot \frac{Precision \cdot Recall}{Precision + Recall} \tag{2}$$

4   Experiments

This section shortly introduces the data sets that were used for the evaluation of the combined software as well as the new feature selection procedure.

4.1   Data

The real world data set contains billing information from German sickness funds that are stored in a anonymized data base of Gesundheitsforen Leipzig (FDB).

They cover diagnoses, medication, demographic and cost information. For example all patients with a risk or diagnosis of cerebral infarction (ICD-Code I63) are contained in this set.

Cerebral Infarction Patients Data Set. To analyze the behavior of the dmt4sp algorithm regarding run time and number of produced episode rules a subset of patients diagnosed with I63 in hospital was extracted from the FDB. The chosen patients were covered in the database at least from 2008 to 2012, receiving their first inpatient I63 diagnosis in 2012. This resulted in 2178 patient histories spanning five to six years which is the longest time span that can be found in the database. The event types included in the histories are drug prescriptions and diagnoses. The analyses were performed on six different time step granularities: day, week, month, quarter, half-year and year. During one time step any event type can occur once at most, i.e. larger time steps lead to a reduction of the number of events. In the test data set this corresponds to the following amount of events for each granularity:

Histories	Event types	Day	Week	Month	Quarter	Half-year	Year
2178	2775	991,738	990,037	971,619	907,782	614,165	425,181

Cerebral Infarction High Risk Patients Data Set. For the feature selection procedure a data set of patients with especially high risk of target event I63 was defined: Patients covered from 2008 to 2012 and a I63 diagnoses no earlier than 2010, with additional diagnoses of risk factors diabetes mellitus (E10-E14), hypertensive diseases (I10-I15) and peripheral artery occlusive disease (I73.9). After splitting the data set by the target event, the positive class (C^+) contains 645 patient histories, the negative class (C^-) 13958. Thus the relative frequency of cerebral infarction in the high risk data set is 4.62 %. Event types included are drug prescriptions (ATC-code on fourth level; prescription day), inpatient diagnoses (ICD-code on third level; hospital discharge day) and outpatient diagnoses (ICD-code on third level; last day of quarter with diagnose).

4.2   Run Time Analysis of Extended Lifelines2 Version

The first contribution of this paper is the inclusion of an episode miner into Lifelines2. To show the performance of episode mining within the Lifelines2 framework, we conducted several experiments on real world data under various constraints. First of all, we explored how dmt4sp runs, when different time step intervals (granularities) are used (cf. Fig. 4). Analyzing the performance of the dmt4sp algorithm showed, that changing the minimum confidence has next to no influence on the run time. In Fig. 4c this is shown exemplarily at time step granularity "year". Therefore a medium confidence (0.5) was chosen to directly compare the algorithm run times for different granularities. As can be seen in Fig. 4b the coarser the granularity the shorter the run time, as was to be expected

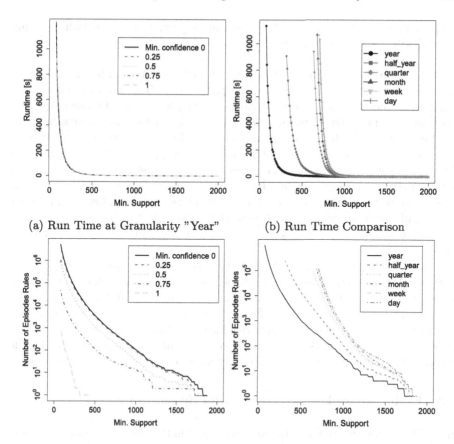

(a) Run Time at Granularity "Year"   (b) Run Time Comparison

Fig. 4. Analysis of the run time of dmt4sp algorithm and the number of created episode rules depending on the minimum support, minimum confidence and time step granularity in the cerebral infarction data set. As shown exemplarily in (a) minimum confidence has almost no influence on the run time. Therefore only a comparison of the time granularities at confidence minimum 0.5 is shown in (b). In (c) the number of episode rules at time step granularity "year" is shown and in (d) comparison at minimum confidence 0.5. Episodes found at a high minimum support will also be included at any lower minimum support. The graph is therefore partly cumulative For each granularity the analysis was terminated after first reaching run times longer than 15 min.

from the decreasing number of events. Moreover, run times for the granularities day to quarter are fairly similar since the majority of events (outpatient diagnosis) is given at quarter intervals. In all cases the run time increases strongly exponential with decreasing minimum support. However the start of the sudden run time increase varies greatly for the larger time steps. Since the increase in run time is very sharp it might be difficult to find a minimum support that is low enough for the specific research purpose without reaching unfeasible run times. With the 15 min threshold on this test data set the lowest reachable minimum support values were ~ 3 % for the granularity year, ~ 14 % for half-year and around 30 % for the finer granularities.

Results Number of Episodes. The analysis of the number of produced episodes showed that a high minimum support in combination with a low minimum confidence produces the most rules (see Fig. 4c). Many of these rules however, have no predictive qualities. On the other hand low minimum support in combination with high minimum confidence produces very few rules. Of course, all episode rules found at a very strict minimum support (high values) are also found at lower minimum support values. Most of these rules apply to only a small minority of the data, but have very good predictive qualities. We are mostly interested in the qualities of the rule set as a whole. Therefore, a larger number of very specific rules, that as a whole cover most of the patients, is just as valid as a few generic, but also very precise rules, which are usually hard to find.

Similarly to the run time analysis the time granularities quarter to day behave fairly similar with regard to the number of found episode rules. Moreover, dmt4sp runs with the granularities year and half-year produce far less episode rules than with the finer granularities, for any tested minimum support value. This was to be expected since these histories have less time steps and less events to build episode rules from (see Fig. 4d). Throughout all time step granularities the number of episode rules increases strongly with decreasing minimum support.

4.3   Evaluation of Feature Granularity Selection Process

Run Time Analysis Results. In our run time analysis of the feature selection procedure we determined that the episode mining step is up to ten times faster than the determination of confidences (see Fig. 5c). This validates our efforts to accelerate the latter step through introduction of parallelization. As can also be seen in Fig. 5c, this endeavor was very successful, since run times are now up to four times shorter than before. Additionally, we found that run times become considerably longer the more general the involved event types are. The analysis with an event hierarchy that was only expanded once took over 40 times as long as the analysis with the three times expanded hierarchy. Which in turn, justifies our efforts to keep events as specific as possible during feature selection.

F-measure Analysis. The execution of the feature selection procedure for the four different class ratios took roughly 12 hours for each combination. Although the increase in $|C_-|$ lead to slightly longer run times for each run. The F1-measure at iteration zero shows the performance of the classifier without feature selection. In all cases feature selection did improve the final F1-measure of the learned episode rule set (see Fig. 5b). In all but one case improvements did not begin before iteration ~ 140. This shows that the collapse of the most detailed event levels does usually not improve the F1-measure. Only the collapse of less specif event levels will bring improvements. We assume that additional iterations would have improved the F1-measure further. Figure 5b also shows that the class

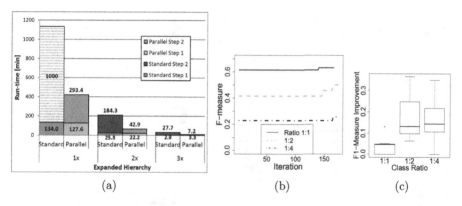

(a)   (b)   (c)

Fig. 5. Figure (a) shows the difference in run time before (red) and after (green) the introduction of parallelization to the feature selection process at different expansion states of the event hierarchy. The initial hierarchy used was expanded to first level for all event types, aside from the target event. Expanding once more translates in most cases to the second level, expanding twice to the third level and expanding thrice to the most detailed level for ICDs. Only ATC codes can be expanded one more time. The steps mentioned correspond to the iteration steps shown in Fig. 3. The run time for step 2 with once expanded hierarchy under standard conditions (shaded area) has been approximated. Figure (b) shows that feature selection improves the predictive quality of the mined episode rules, but only in later iterations. The F1-measure at iteration zero shows the performance of the classifier without feature selection. Skewed class ratios of C_+ and C_- decrease the overall quality. We performed bootstrapping ten times for every ratio and show the resulting overall improvements in (c) (Color figure online).

ratio influences the F1-measure considerably. It is possible that this a data set specific problem, since we believe that the prediction of cerebral infarction is very difficult, if possible at all with this method. Since we sampled only a subset if the negative class we performed bootstrapping ten times for each ratio and analyzed the resulting F1-measure improvements. Since our feature selection algorithm is greedy, it is possible that no improvements are made at all. This happens more often at class ratio 1:1 than at the more skewed class ratios, since the original F1-measure is relatively good already. Overall, the improvement made by the feature selection is larger, where the original F1-measure is smaller.

Real World Data Set Analysis. We exemplarily discuss the impact of feature selection on a subset of 200 patient histories of the high risk data set (class ratio 1:1). In 24 min 170 iterations were performed. Starting from an F-measure of 0.57, feature selection improved the quality of the mined episode rule set to 0.73. All improving event collapses in this run were medication codes, suggesting that instead of using the chemical substance hierarchy level, it is beneficial to use a less detailed level. Figure 6 shows not only the incorporation of the episode rule mining into the software, but also how the results of the feature selection process are displayed.

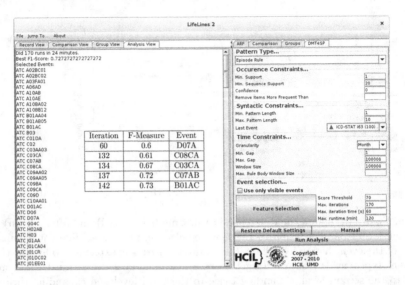

Fig. 6. The new episode mining interface after a feature selection run. The left side shows a summary of the results, including number of iterations, run time, final F-measure and selected event types. The right side shows the entry form for the configuration of dmt4sp constraints and feature selection options. The table shows which event collapse improved the current F-measure, but is not part of the screen-shot.

5 Conclusion

This paper brings two merits for episode rule mining (ERM): First, we merged an ERM software (dmt4sp) with a visualization tool (LifeLines2), to make the exploration of timed event sequences and their pattern easier. Second, we introduce a feature granularity selection procedure for hierarchical data sets that often arise in medical domains. This procedure traverses the space of hierarchical events in a bottom up manner in order to identify the best set of event-granularity. This traversal respects time, minimum support and rule-quality constraints. The experiments on real world data show that the inclusion of dmt4sp into LifeLines2 enables the user to do a guided episode mining. Moreover, the automated feature selection algorithm is able to find a locally optimal event set so that it is not necessary to define the respective events up-front. For future work, we would like to improve the feature selection technique by using heuristics. Then, the choice of granularity should result quicker in an optimal event set. Finally, we believe that the combination of episode visualization and automated pattern extraction may lead the user quicker to interesting insights to their data.

References

1. Mannila, H., Toivonen, H., Verkamo, A.I.: Discovery of frequent episodes in event sequences. Data Min. Knowl. Discov. **1**(3), 259–289 (1997)

2. Zimmermann, A.: Understanding episode mining techniques: benchmarking on diverse, realistic, artificial data. Intell. Data Anal. **18**(5), 761–791 (2014)
3. Rigotti, C.: Dmt4sp (2014). http://liris.cnrs.fr/~crigotti/dmt4sp
4. Brownstein, J., Freifeld, C., et al.: Healthmap: the development of automated real-time internet surveillance for epidemic intelligence. Euro Surveill. **12**(11), E071129 (2007)
5. Duke, J.D., Li, X., Grannis, S.J.: Data visualization speeds review of potential adverse drug events in patients on multiple medications. J. Biomed. Inf. **43**(2), 326–331 (2010)
6. Plaisant, C., Milash, B., Rose, A., Widoff, S., Shneiderman, B.: Lifelines: visualizing personal histories. In: Proceedings of the SIGCHI Conference on Human Factors in Computing Systems, pp. 221–227. ACM (1996)
7. Pieczkiewicz, D.S., Finkelstein, S.M., Hertz, M.I.: Design and evaluation of a web-based interactive visualization system for lung transplant home monitoring data. In: AMIA Annual Symposium Proceedings, vol. 2007, p. 598. American Medical Informatics Association (2007)
8. Wongsuphasawat, K., Shneiderman, B.: Finding comparable temporal categorical records: a similarity measure with an interactive visualization. In: IEEE Symposium on Visual Analytics Science and Technology, VAST 2009, pp. 27–34. IEEE (2009)
9. Klimov, D., Shahar, Y., Taieb-Maimon, M.: Intelligent selection and retrieval of multiple time-oriented records. J. Intell. Inf. Syst. **35**(2), 261–300 (2010)
10. Han, J., Kamber, M.: Data Mining: Concepts and Techniques. Morgan Kaufmann, San Francisco (2006)
11. Jain, L.C., Holmes, D.E.: Data Mining: Foundations and Intelligent Paradigms. In: Holmes, D.E., Jain, L.C. (eds.) Medical, Health, Social, Biological and other Applications, vol. 3. Springer, Heidelberg (2012)
12. Rao, R.B., Krishnan, S., Niculescu, R.S.: Data mining for improved cardiac care. SIGKDD Explor. Newsl. **8**(1), 3–10 (2006)
13. Wang, T.D., Plaisant, C., Shneiderman, B.: Visual information seeking in multiple electronic health records: design recommendations and a process model. In: Proceedings of the 1st ACM International Health Informatics Symposium, pp. 46–55. ACM (2010)
14. Wang, T.D., Deshpande, A., Shneiderman, B.: A temporal pattern search algorithm for personal history event visualization. IEEE Trans. Knowl. Data Eng. **24**(5), 799–812 (2012)

An Unexpectedness-Augmented Utility Model for Making Serendipitous Recommendation

Qianru Zheng[1], Chi-Kong Chan[2], and Horace H.S. Ip[1(✉)]

[1] Department of Computer Science, City University of Hong Kong,
Kowloon, Hong Kong
qrzheng2-c@my.cityu.edu.hk, cship@cityu.edu.hk
[2] Department of Computing, Hang Seng Management College, Shatin, Hong Kong
chanck@hsmc.edu.hk

Abstract. Many recommendation systems traditionally focus on improving accuracy, while other aspects of recommendation quality are often overlooked, such as serendipity. Intuitively, a serendipitous recommendation is one that provides a pleasant surprise, which means that a suggestion must be unexpected to the user, and yet it must be useful. Based on this principle, we propose a novel serendipity-oriented recommendation mechanism. To model unexpectedness, we combine the concepts of item rareness and dis-similarity: the less popular is an item and the further is its distance from a user's profile, the more unexpected it is assumed to be. To model usefulness, we adopt PureSVD latent factor model, whose effectiveness in capturing user interests has been demonstrated. The effectiveness of our mechanism has been experimentally evaluated based on popular benchmark datasets and the results are encouraging: our approach produced superior results in terms of serendipity, and also leads in terms of accuracy and diversity.

Keywords: Serendipity · Diversity · Recommendation systems

1 Introduction

Recommendation System (RS) has become a vital part of e-commerce websites. To the users, it provides useful and personalized product recommendations. To the merchants, it provides an effective cross-selling solution. Because of its usefulness, RS has been successfully applied to various application areas, ranging from traditional applications such as movies recommendation [19] in Movielens and Netflix, products recommendation in Amazon [15] and book recommendations [25], to more recent applications in tourism and travel recommendation [16], and social network recommendation [12].

Traditionally, RS algorithms aim at improving recommendation accuracy (e.g., root mean square error (RMSE)), and particularly, recommendation precision. In both collaborative filtering (CF) and content based (CB) methods, precision measures the proportion of the recommended items that are chosen by

© Springer International Publishing Switzerland 2015
P. Perner (Ed.): ICDM 2015, LNAI 9165, pp. 216–230, 2015.
DOI: 10.1007/978-3-319-20910-4_16

a user (called the hit items). In order to maintain good accuracies, many recommendation systems tend to recommend only items that are relevant and similar to the user's previous choices, i.e. those items that match the user's profile. After all, such kind of recommendations is intuitive, safe and usually accurate. However, an over-emphasis on accuracy may restrict the users' choices to the items most similar to his/her previous selections. After all, a user may get bored of the usual item genres, and may want a recommendation off the beaten path. Moreover, some recommendation may simply be too obvious that the user can find it himself even without recommendations. Consider, for example, recommending yet another *Harry Potter* series movie to someone who has already owned a full set of it. The effectiveness of such suggestions is questionable.

To handle this issue, other aspects of recommendation quality should also be taken into account. Indeed, a number of alternative approaches have been proposed in recent years; among them are novelty, diversity and serendipity. The novelty-based (distance-based novelty) approaches consider the newness of the item from the users' prospective. In practice, this is often modeled as the level of dissimilarly (i.e., the distance) between an item and the user's profile. The diversity-based approaches, as the name suggests, aim at providing a diversified list of recommendations to the users. Two main approaches were proposed, namely intra-list diversity, which deals with diversity within a list of recommended items, and aggregate diversity, which deals with the overall diversity across all users. Both novelty and diversity can provide recommendations outside the usual item genres that are previously favored by the user. However, one can argue that either approach, when working on its own, may not necessarily lead to useful recommendation. In light of this, a number of researches have turned to the concept of serendipity. Intuitively, a serendipitous event is one that will result in a pleasant surprise. Serendipitous recommendation algorithms thus aim at providing items which are both unexpected and useful to the users [8]. A good serendipitous recommendation system not only broadens the user's choices (since serendipitous recommendations do not restrict themselves to items similar to the user profiles or the popular items), but also provides a valuable tool for e-retailers to cross-sell their off-the-beaten-track products as well.

In this paper, we propose a scheme for making serendipitous recommendations that are both unexpected and useful to users. First, in order to model unexpectedness, two factors are considered, namely, item rareness and item dissimilarity from the user profile. The rationales are as follows. Recommending popular items will result in low unexpectedness because these items are likely to be well known to the users already, and therefore would bring little surprises even if they may fit a user's profile. Similarly, recommending items that are similar to a user's profiles may result in items already familiar to the user (for instance, a sequel to a user's favorite movie). In both cases, it is likely that the user can find the items easily even without recommendation system. In contrast, less popular items or the items not similar to the user profile would provide higher level of unexpectedness to the user.

Yet, unexpectedness alone is still not sufficient to make the user feel serendipitous. To achieve this goal, one also need to ensure that the recommendations are useful and favored by the user. To model usefulness, we adopted a PureSVD model. PureSVD is a latent-factor-model based collaborative filtering algorithm that is able to provide high quality recommendation. In this work, the scores for unexpectedness are introduced into the utility model, which forms the basis of our recommendation. The result is a list of items that are not only unexpected to the user but also useful to them as well.

The contributions of this paper are as follow. Firstly, we propose a novel serendipitous recommendation algorithm by considering both unexpectedness and usefulness of the recommended items. Secondly, we also provided a formal model of unexpectedness based on two factors, namely, item-rareness and an item's distance from the user profile. Finally, we provide detailed experimental comparisons with other well-known serendipity-oriented algorithms as well as other baseline methods. Experiments showed that our proposed scheme achieves superior results not only in terms of serendipity, but also lead in other important metrics as well, including precision, intra-list diversity and aggregate diversity.

The rest of the paper is organized as follows: Sect. 2 reviews the related works and the relevant concepts. Section 3 presents the proposed scheme. Experiment design and results are shown in Sect. 4. Finally, conclusion is given in Sect. 5.

2   Related Works

While many works on recommendation systems focused mainly on improving recommendation accuracy [4,22,23], some researchers [5,8,9,17,18] have argued that accuracy alone is not sufficient in evaluating recommendation quality. Instead, several new concepts have been proposed recently, namely, diversity, unexpectedness and serendipity.

The first related concept is diversity. There are two main approaches, namely, aggregate diversity and intra-list diversity. Intra-list diversity [10,25] refers to the difference between each pair of items in a recommendation list. In [25], Ziegler et al. proposed a scheme for improving the intra-list diversity by diversifying the topic of the recommendations. In [10], Zhang et al. proposed a method which optimizes both accuracy and diversity. However, intra-list diversity does not necessarily give rise to serendipity. For example, providing a user with a list of movies of various genres (e.g. animation, adventure and action) would certainly increase the intra-list diversity. Yet, such a recommendation could still be similar to the user's previous choices if the user has watched all these types of movies. Such a list would still have high diversity and reasonable accuracy, but would not surprise the user. Different from intra-list diversity, aggregate diversity [2,3] measures the total number of distinct recommended items across all users. For example, Adomavicius et al. [3] argued that many recommendation algorithms tend to have a bias toward the more popular items because those items have more historical data (i.e., user ratings) and hence they would be recommended more frequently. As pointed out by the authors, such kind of recommendations would reduce the aggregate diversity because the same pieces of popular items would tend to be recommended to multiple users. To solve this issue, several re-ranking methods have been proposed. This includes, for example, a method that ranks the recommendations, where the predicted ratings are higher than a certain threshold, in reverse to their popularity. It was then argued that a high aggregate diversity could help to expand the user's horizon because the recommendations would not be restricted to the popular items. Moreover, it would also be beneficial to the merchants because they can profit from not only the popular items but also from the 'long-tail' items (the items located in the tail of the sales distribution). However, despite the claimed benefits, aggregate diversity is not a replacement for serendipity. A list of recommendations with high aggregate diversity, which provides many distinct items across a large group of users, does not necessarily provide items that are both unexpected and useful to the individuals. Moreover, aggregate diversity is a measurement calculated across from all users, while the serendipity is calculated for individual users.

Another related concept is unexpectedness. In [1], Adamopoulos et al. summarized various proposed definitions of unexpectedness, including associating unexpectedness to the prior background knowledge of decision makers [20], and measuring unexpectedness by taking multi-facets (frequent itemsets, tiles, association rule and classification rule) into account. In [1], Adamopoulos et al. argued that unexpectedness should consider the expectation of users, where unexpectedness is obtained by generating the recommendation significantly depart from the user expectedness. In [8], Ge et al. discussed that unexpected recommendation could be viewed as the recommendations which do not belong to the primitive prediction model. Although various definition of unexpectedness are proposed, unexpectedness is not equal to serendipity. According to Ge et al. [8], unexpectedness is one of the most important components of serendipity.

Serendipity differs from diversity and unexpectedness in that it attempts to model the users' level of positive surprise toward the items. Literally speaking, the word serendipity denotes a pleasant surprise, or a fortunate yet unexpected discovery by chance. Thus, a serendipitous discovery should be unexpected, yet useful. This idea has been explored by a number of works. For example, in [8], Ge et al. discussed the idea that serendipity should cover unexpectedness (i.e., items that are not yet discovered and unexpected by the user) and usefulness (i.e., items that are of interest to the user). However, unexpectedness and usefulness are not clearly defined in this work. Some other works have further elaborated the definition of unexpectedness and usefulness. In [1], Adamopoulos et al. adopted the definition of usefulness of an item that is determined by its average rating: if its average rating among the users is larger than a certain threshold, then it is considered to be useful for all the users. In [21] usefulness of an item is determined by the user's rating for the recommended item. In this work, we follow a definition of usefulness similar to an approach adopted in [21], which provides personalized usefulness estimation and better reflects real life situation. Regarding unexpectedness, it was defined differently in [1,21].

In [1], Adamopoulos et al. proposed that high unexpectedness can be obtained by recommending items that are different from the set of the expected items. In [21], Lu et al. argued that since the popular items are so well-known, they are easy to find, hence they would lead to low unexpectedness. Partially inspired by these works, we argue in this paper that the unexpectedness of an item should be associated with both the item's popularity (or rareness) and the item's level of dissimilarity (i.e., its distance) from the user profile.

3   The Proposed Scheme

In this section, we formally present our proposed scheme based on the ideas developed in the previous section. First, the unexpectedness of an item is defined in Sect. 3.1. After that, item utility is presented in Sect. 3.2. And finally, our optimization process is described in Sect. 3.3.

3.1   Unexpectedness

As explained in Sect. 2, the unexpectedness of an item depends on two factors: the item popularity (or rareness) and the item's dissimilarity from the user profile. Firstly, regarding item popularity, the more popular is an item, the lower is its unexpectedness for a user because such items would be so well known that user could find them easily even without recommendations. This idea is implemented in Eq. 1, where $Pop(i)$ denotes the number of users who have selected item i, and $|U|$ is the number of all users.

$$Unexpectedness(u, i) \propto 1 - \frac{Pop(i)}{|U|} \tag{1}$$

Secondly, regarding an item's dissimilarity from the user profile, the concept is depicted in Eq. 2, where $S(u)$ is the set of items chosen by u (the user), and $diff(i,j)$ denotes the degree of dissimilarity between item i and item j (see below). As seen from Eq. 2, the dissimilarity of an item i to a user u is high if i is different from the other items that s/he has chosen before.

$$Unexpectedness(u, i) \propto \frac{\sum_{j \in S(u)} diff(i,j)}{|S(u)|} \tag{2}$$

The dissimilarity function $diff(i,j)$ can be obtained by $diff(i,j) = 1 - sim(i, j)$, where $sim(i, j)$ denotes the degree of the similarity between i and j. In the literature, there are various possible ways for computing the similarity between a pair of items, including both content dependent [4] and content independent metrics [23]. In this paper we adopt a content independent metric [23] for our similarity function, which is illustrated in Eq. 3, where $S(i, j)$ is the set of users (co-rated users) who have chosen both item i and item j, $r_{u,i}$ is user u's rating for item i, and $\overline{r_u}$ is the average rating of u for his rated items. Basically, Eq. 3 measures the rating consistency among the co-rated users on i and j. If the co-rated users consistently give high (or low) ratings to both i and j, it would indicate that i and j are similar.

$$sim(i,j) = \frac{\sum\limits_{u \in S(i,j)} (r_{u,i} - \overline{r_u}) \cdot (r_{u,j} - \overline{r_u})}{\sqrt{\sum\limits_{u \in S(i,j)} (r_{u,i} - \overline{r_u})^2} \sqrt{\sum\limits_{u \in S(i,j)} (r_{u,j} - \overline{r_u})^2}} \tag{3}$$

Finally, the unexpectedness of an item i to user u is defined as a linear combination of the item's rareness and its dissimilarity from the user's profile:

$$Unexpectedness(u,i) = (1 - \frac{Pop(i)}{|U|}) + \frac{\sum_{j \in S(u)} diff(i,j)}{|S(u)|} \tag{4}$$

3.2 Utility

Recommending an item based solely on unexpectedness may lead to a risk. That is, the items could be too unexpected and deviate too far from the user's interest. In either case, the user's trust and satisfaction for the system would decrease. Hence, in addition to unexpectedness, we must also consider the utility of the recommendation items. Utility measures an item's relevance and usefulness to the user. In practice, utility is usually measured by the predicted rating for an item by a given user. To predict an item's utility, we apply latent factor models which are good at predicting items utility and perform well in capturing user future interests. The model we adopt in this paper is PureSVD [7].

The majority of latent factor models are based on the factorization of the user-item rating matrix by Singular Value Decomposition (SVD). The main idea of SVD models is to factorize the user-item rating matrix into three low rank matrices (Eq. 5). U is $n \times k$ orthonormal matrix, Q is $m \times k$ orthonormal matrix and Σ is $k \times k$ diagonal matrix with the top k singular values. k is the number of latent factors. Alternatively, $\hat{R}$ can be represented by Eq. 6. $\hat{R}$ is the estimated utility matrix.

$$\hat{R} = U \cdot \Sigma \cdot Q^T \tag{5}$$

$$\hat{R} = P \cdot Q^T \tag{6}$$

After factorization, each user is associated with a k-d vector p_u, representing the user u's preference for k factors. And each item is also associated with a k-d vector q_i, describing i's importance weight for k factors. The number of latent factors (k) is 50 in this paper. PureSVD is a standard latent factor model that measures the utility between i (the item) and u (the user) by the product of user-factor vector p_u and item-factor vector q_i (Eq. 7).

$$Utility(u,i) = p_u \cdot q_i{}^T \tag{7}$$

3.3   Optimization

In recommendation systems, the utility of an item refers to the attractiveness of the item to a user. In practice, this is often estimated based on the observed user-item ratings. In order to predict the ratings accurately, a model must first be trained based on known historical data. A typical approach is shown in Eq. 8. Here, $r(u, i)$ is the observed rating of user u for item i, and $\lambda(\|p_u\|^2 + \|q_i\|^2)$ is a regularizing term to prevent overfitting.

$$min \sum_u \sum_{i \in S(u)} (r(u, i) - p_u \cdot q_i^T)^2 + \lambda(\|p_u\|^2 + \|q_i\|^2) \qquad (8)$$

Equation 8 illustrates the traditional approach, where unexpectedness is not considered. In order to take serendipity into account, we can employ a weight $w_{ui} = Unexpectedness(u, i)$ for penalizing items that are popular and similar to the user's profile. Also, note that Eq. 8 only optimizes the errors on the observed items ($i \in S(u)$). However, as pointed out by [24], both the unobserved and observed items contribute to recommendation accuracy (e.g., the top n recommendation accuracy). In light of this, Eq. 8 has been readapted accordingly to include all items. The revised version is shown in Eq. 9 and the corresponding learning process is depicted in Algorithm 1, where γ is the learning rate. For distinguish purpose, we used $\tilde{r}(u, i)$ instead of $r(u, i)$, where $\tilde{r}(u, i)$ represents both observed and unobserved ratings. Our proposed method is simple to implement, and can easily be applied to real life e-commerce systems. Most importantly, experimental results indicate that the proposal method performs well in both accuracy and diversity. The detailed findings will be presented in the next section.

$$min \sum_u \sum_{i \in I} (\tilde{r}(u, i) - p_u \cdot q_i^T)^2 \cdot w_{ui} + \lambda(\|p_u\|^2 + \|q_i\|^2) \qquad (9)$$

Algorithm 1. Update of p_u and q_i

---
for $u \in U$ **do**
    for $i \in I$ **do**
        $err(u, i) = (\tilde{r}(u, i) - p_u \cdot q_i^T) \cdot w_{ui}$
        $p_u \leftarrow p_u + \gamma(err(u, i) \cdot q_i - \lambda \cdot p_u)$
        $q_i \leftarrow q_i + \gamma(err(u, i) \cdot p_u - \lambda \cdot q_i)$
    end for
end for

---

4   Experiment

To evaluate our proposed method, we conducted a series of experiments on two representative datasets. In Sect. 4.1, the adopted datasets are first introduced. Experiment setup is presented in Sect. 4.2. Finally, experiment results for our scheme as well as those of representative approaches are discussed in Sect. 4.3.

4.1   Datasets

Two representative datasets were chosen to evaluate our proposed scheme, namely, Netflix [11] and Movielens [6]. Both datasets contain user rating data collected over long periods and they are both widely used for evaluation in the literature. There are 2,113 users, 10,197 items and more than 800k ratings in Movielens dataset, dating from October 1997 to December 2008. Its sparsity is about 3.976 %. The original Netflix data set contains over 17k items, 480k users and 100M ratings dated from 1997 to 2008. For the sake of scalability and comparability, we randomly sampled the ratings from 2000 users from the original Netflix dataset. The resulting dataset contains 5,260 items and 632,335 ratings. The statistical properties of these two datasets are summarized in Table 1.

Table 1. Statistical properties of two datasets

	# of users	# of items	# of ratings	Sparsity
Movielens	2,113	10,197	800k	3.976 %
Netflix	2,000	5,260	635k	6.01 %

4.2   Experiment Setup

Each dataset was split into two disjoint sets chronologically, with the older data in the training set and the remaining data in the test set. Recommendations were generated based on the training set. Each user was provided with 10 lists of recommendations, with size of 10, 20,..., and 100 items respectively. The value of α is 0.5.

A number of metrics were employed to evaluate the recommendation quality. The first one was accuracy. Two accuracy metrics were adopted, namely, precision and recall, which are defined by Eqs. 10 and 11. Here, $RS(u, N)$ represents the top N recommendations in the recommendation list of user u. $TestSet(u)$ is the set of items in the test set that are chosen by user u. The precision metric measures the proportion of recommendations among the recommendation list which are actually selected by the users (the proportion of hit items). Recall measures the proportion of the recommendations which are actually selected by the users among the items relevant to the users.

$$Prec@N = \frac{\sum_u |RS(u, N) \cap TestSet(u)|}{N \cdot |U|} \tag{10}$$

$$Recall@N = \frac{\sum_u |RS(u, N) \cap TestSet(u)|}{|TestSet(u)| \cdot |U|} \tag{11}$$

The second evaluation metric is serendipity (Eq. 13). Later in this section, we shall present our experimental evaluation results alongside with a number of representative approaches, including two other serendipity based models. In order to provide fair and meaningful comparisons, we decided to adopt the top N

serendipity metric that has also been utilized by these benchmark approaches for evaluation purpose [1,8,21]. The top N serendipity metric is in some way similar to precision, except that it is stricter. In precision, one only counts the number of hit items in a recommendation list. In a serendipity-oriented metric such as the top N serendipity, on the other hand, one also needs to determine whether the hit items are unexpected and useful for the user. Recall from previous section that serendipity depends on two factors, namely unexpectedness and usefulness. To evaluate unexpectedness, a model (Predictive Model (PM)) consisting of a set of items which are assumed to be expected for the users is first constructed. And any recommended items that are not included in the set of recommendations generated by the Predictive Model (PM) is treated as the unexpected ones. The concept is illustrated in Eq. 12. Following the practice of [1,21], the set of expected items generated by PM consists of 100 items, which includes the top 50 items with highest average rating and the top 50 items with highest popularity value. To measure the usefulness of the recommendations, we observe whether the user selects the recommended item and favors it (i.e., gives it a high rating). In this metric, the set of high-rating-items are those items with rating larger than a given threshold θ. The set of useful items is then defined as $USEFUL(u) = \{i \in TestSet(u)|r(u,i) > \theta\}$, where θ is the threshold rating. In our experiment, the ratings of two adopted datasets have a range of zero to five, and θ's value is 3.

$$UNEXP(u,N) = RS(u,N)\backslash PM \tag{12}$$

$$SRDP@N(u) = \sum_u \frac{|UNEXP(u,N) \cap USEFUL(u)|}{N \cdot |U|} \times 100\% \tag{13}$$

The third metric is intra-list diversity. The calculation is shown in (Eq. 14), where $diff(i,j)$ is the dissimilarity between item i and item j. We adopted a content-independent metric [10] to calculate the (dis)-similarity between any two items, which is illustrated by Eq. 4. The difference function $diff(i,j)$ is then obtained by $1 - sim(i,j)$. (A point of note, the dissimilarity in intra-list diversity is not the same as the dissimilarity that is used to compute unexpectedness. Intra-list diversity measures the difference between each pair of items in the recommendation list, while unexpectedness concerns the difference between the candidate recommended item and the user's previous chosen items.)

$$IntralistDiversity@N = \frac{1}{|U| \cdot N(N-1)} \cdot$$
$$\sum_u \sum_{i \in RS(u,N)} \sum_{j \neq i \in RS(u,N)} diff(i,j) \tag{14}$$

A related metric is the aggregate diversity [3], which measures the number of distinct items recommended across all users (Eq. 15). A high aggregate diversity in recommendation is beneficial to the e-retailer since it indicates that more distinct items are recommended to the users, thus increases the sale potential.

$$AggDiversity@N = |\cup_{u \in U} RS(u,N)| \tag{15}$$

To evaluate our scheme, we compared the performance of our method with a number of representative methods. Four schemes for making personalized recommendations are included in this study, including two latent factors models, namely SVD (bias) and SVDpp [7,13,14] and two serendipitous recommendation algorithms, namely Adamopoulos's method [1] and Lu's method [21]. The latent factor based models have gained a lot of attentions in RS because of their significant performance in top N recommendations. Adamopoulos's method and Lu's method are both two representative serendipitous recommendation algorithms, whose performance in making serendipitous recommendations has been demonstrated. Apart from these representative methods, we also implemented three other non-personalized approaches to serve as benchmark algorithms, namely, AvgRating, Random and Toppop. AvgRating recommends the items which have the highest average ratings to the user. Random uses a random algorithm to recommends the non-chosen items to the users. Toppop recommends the most popular items to the users.

4.3   Experiment Results

Accuracy Performance. In this section, we will show the comparison of our method and other baseline methods in top n accuracy. Figure 1 shows the top N precision of various methods on Movielens and Netflix datasets, Fig. 2 shows the recall. The performance of Random and Avgrating turned out to be very close in this case. Thus, for clarity, only the Random method is shown.

Several observations can be made. Firstly, all personalized algorithms performed better than the non-personalized benchmark methods. Secondly, among the personalized methods, latent-factor-based methods (which include our method, Lu's method, SVDpp and SVD(bias)) produced the best performance. (A side note, interested readers may refer to [7] for a detailed discussion on the effectiveness of the latent factor models). Thirdly and most importantly, our method performed the best on both datasets. For example, in Movielens dataset, our method achieved a 10 % improvement in precision over the second best method (Lu's method), and up to 50 % better than the third best method (SVD (bias)) for $N = 10$. Similar results can be observed from Netflix dataset. Results of other metrics also support similar conclusions. We attribute the good top n accuracy performance of our method to two reasons. The first one is the adopted utility model-PureSVD. It is reported that PureSVD performs well in top n accuracy [7]. The second reason is that, as mentioned in Sect. 3.3, our method explicitly models both observed and unobserved data. According to the work of Steck [24], in applications where the data are not *missing at random* (in the context of recommendation systems, *missing at random* means that the probability of a rating to be missing does not depend on its value), the missing data may contain hidden implications (e.g., many users simply do not provide ratings for the movies they do not like). Such missing ratings have to be modeled as to obtain better results, and they are included in our model for this reason.

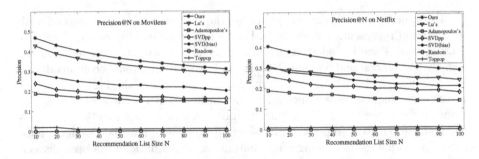

Fig. 1. Comparison of precision on Movielens and Netflix datasets

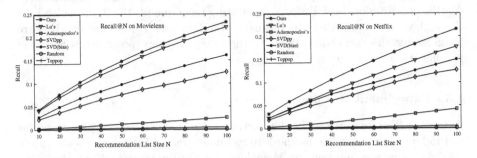

Fig. 2. Comparison of recall on Movielens and Netflix datasets

Serendipity Performance. Next, we evaluated the serendipity performance of the various methods using top N serendipity (Eq. 13). The results are shown in Fig. 3. Once again the results of Random and Avgrating are very close to each other, so for clarity, only the result for Random is shown in the figure.

From Fig. 3, we observe that our method outperforms all other methods on the two datasets in serendipity significantly. For example, in the Movielens dataset, for $N = 100$, our method led the second best method (Lu's method [21]) by 32 %, and third best method (SVD(bias)) by 60 %. For smaller Ns, the difference is even larger (for instance, our score for $N = 10$ is 2.57 times of that of the second highest method). Similar results can be observed from Netflix dataset. For example, when $N = 50$, our performance was 76 % higher than the second best method (SVD(bias)). This result is very encouraging because normally one would expect a method that does well in precision would not necessarily achieve high scores in serendipity. The reason is that by offering off-the-beaten-track recommendations, one would think that the precision would suffer because the most popular and straightforward (and hence "safe") recommendations are now excluded. Yet, our results seem to suggest that it is not necessarily the case, and it is possible to achieve good precision and good serendipity by considering both utility and unexpectedness in the framework.

Additionally, several other observations can be made from Fig. 3. Firstly, as in precision, personalized methods performed better than non-personalized

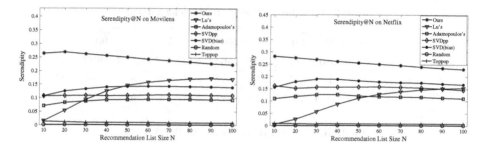

Fig. 3. Comparison of serendipity on Movielens and Netflix datasets

methods. Secondly, among the personalized methods, it is interesting to note that the two non-serendipitous methods (SVD(bias) and SVDpp) actually performed quite well in serendipity (for example, in the Netflix dataset, the two non-serendipitous methods actually outperformed the remaining two serendipity-oriented approaches for all $N < 90$, although this was not the case in the Movielens dataset). Thirdly, in all approaches except Lu's method [21], the top N serendipity scores were quite stable for different values of N, whereas in Lu's method, it started at a low value but increased as the recommendation list size grew. The reason may be that in Lu's method, the recommendation lists mainly consist of popular or highly-rated items when N is small, which led to low unexpectedness values and resulted in low serendipity. Overall, our method has produced the best performance for all list sizes.

Diversity Performance. Finally, we measured the diversity performance of the various methods. In previous studies, it has been suggested that diversity is achieved at the expense of accuracy [3,25]. However, we argue that accuracy should be the premise of diversity. A list of recommendations achieving high diversity but low accuracy would indicate that recommendations are diverse but do not fit the user's preference. For this reason, we only further evaluated the diversities of the three methods that produced highest accuracy, namely our method, Lu's method [21] and SVD(bias). The results for both intra-list and aggregate diversity are shown in Tables 2, 3, 4 and 5.

From the results, we see that our method achieved the highest diversity (both intra-list and aggregate) among the methods that produced the highest accuracy. For example, for intra-list diversity, our method outperformed Lu's method by 14 % and SVD(bias) by 77 % on the Movielens dataset when N is 10 (Table 2). For aggregate diversity, we obtained values that were more than three times of those obtained by the other two methods for both Netflix dataset (Table 5) and Movielens dataset (Table 4) when $N = 100$. The difference was even more significant for smaller Ns. This is quite remarkable as our method is not primarily designed to improve diversity. The good results can be explained as follows. Recall that our model for unexpectedness consists of two components, namely item (un)popularity and dissimilarity from the user profile. According to

Table 2. Intra list diversity on Movielens dataset

N	10	20	30	40	50	60	70	80	90	100
Ours	**.768**	**.807**	**.831**	**.847**	**.861**	**.871**	**.879**	**.887**	**.893**	**.899**
Lu's method	.674	.744	.779	.812	.832	.846	.861	.870	.880	.887
SVD(bias)	.434	.479	.501	.520	.539	.555	.569	.583	.593	.602

Table 3. Intra list diversity on Netflix dataset

N	10	20	30	40	50	60	70	80	90	100
Ours	**.909**	**.933**	**.945**	**.954**	**.961**	**.966**	**.970**	**.974**	**.977**	**.980**
Lu's method	.881	.909	.918	.930	.940	.950	.958	.964	.967	.970
SVD(bias)	.474	.551	.613	.650	.669	.686	.700	.714	.726	.737

Table 4. Aggregate diversity on Movielens dataset

N	10	20	30	40	50	60	70	80	90	100
Ours	**769**	**1036**	**1244**	**1385**	**1511**	**1637**	**1755**	**1865**	**1961**	**2049**
Lu's method	135	200	257	308	351	395	435	468	503	535
SVD(bias)	83	124	155	180	206	237	258	280	299	321

Table 5. Aggregate list diversity on Netflix dataset

N	10	20	30	40	50	60	70	80	90	100
Ours	**656**	**856**	**985**	**1103**	**1187**	**1259**	**1339**	**1406**	**1468**	**1533**
Lu's method	88	138	177	216	250	284	314	342	374	404
SVD(bias)	66	102	140	174	208	237	259	282	313	338

the work of Adomavicius [3], recommending less popular items results in higher aggregate diversity, which helps to explain our results in Tables 4 and 5. Also, regarding intra-list diversity, recommending the items different from the user profile means that system provides more diverse recommendations that are not restricted to items similar to the user profile. As a result, the intra-list diversity increases. Overall, the results suggest that we have a new approach for providing accurate, serendipitous, and diverse recommendations.

5   Conclusions

In this paper, we proposed a recommendation scheme based on serendipity. There are two requirements for a serendipitous recommendation, namely, that the items must be unexpected, and that the items must be useful to the user. There are two elements that constitute unexpectedness in our model. The first element is item rareness. It is likely that popular items are already well-known to the users, who can find them easily even without recommendation. The second element is the level of dissimilarity to the user-profile, as recommending items that are similar to a user's profiles may also result in items already familiar to the user (for instance, a sequel to a user's favorite movie). Usefulness (or utility) refers to the level of attractiveness of an item to a user. In practice, usefulness is often measured indirectly by the predicted user-item-ratings. In this paper, item utility is modeled using a PureSVD latent factor model, which has been demonstrated to perform well in capturing user future interests. The unexpectedness value and the utility are then combined to obtain a serendipitous score of an item. To evaluate the proposed scheme, its performance is compared with two representative serendipitous algorithms and two popular latent factor models using popular benchmark datasets. The obtained results are very encouraging. Experiment results suggested that, our scheme not only achieved the best performance in terms of serendipity, but it also performed well in term of precision and diversity also. This is significant because it has been previously suggested that serendipity and diversity are achieved at a price of accuracy. Our results seem to suggest a new and effective direction in serendipity-oriented recommendation.

References

1. Adamopoulos, P., Tuzhilin, A.: On unexpectedness in recommender systems: or how to expect the unexpected. In: Proceedings of RecSys 2011, Chicago, IL, USA, 23–27 October 2011, pp. 11–18. ACM, New York (2011)
2. Adomavicius, G., Kwon, Y.: Toward more diverse recommendations: item reranking methods for recommender systems. In: Proceedings of WITS 2009, Phoenix, AZ, USA, 14–15 December 2009. SSRN, New York (2009)
3. Adomavicius, G., Kwon, Y.: Improving aggregate recommendation diversity using ranking-based techniques. IEEE Trans. Knowl. Data Eng. **24**(5), 896–911 (2012)
4. Balabanović, M., Shoham, Y.: Fab: content-based, collaborative recommendation. ACM Commun. **40**(3), 66–72 (1997)
5. Benjamin, W., Chandrasegaran, S., Ramanujan, D., Elmqvist, N., Vishwanathan, S., Ramani, K.: Juxtapoze: supporting serendipity and creative expression in clipart compositions. In: Proceedings of CHI 2014, Tronto, Canada, 26 April–1 May 2014, pp. 341–350. ACM, New York (2014)
6. Cantador, I., Brusilovsky, P., Kuflik, T.: 2nd workshop on information heterogeneity and fusion in recommender systems. In: Proceedings of RecSys 2011, Chicago, IL, USA, 23–27 October 2011. ACM, New York (2011)
7. Cremonesi, P., Koren, Y., Turrin, R.: Performance of recommender algorithms on Top-N recommendation tasks. In: Proceedings of RecSys 2010, Barcelona, Spain, 26–30 September 2010, pp. 39–46. ACM, New York (2010)
8. Ge, M., Delgado-Battenfeld, C., Jannach, D.: Beyond accuracy: evaluating recommender systems by coverage and serendipity. In: Proceedings of RecSys 2010, Barcelona, Spain, 26–30 September 2010, pp. 257–260. ACM, New York (2010)
9. Herlocker, J.L., Konstan, J.A., Terveen, L.G., Riedl, J.T.: Evaluating collaborative filtering recommender systems. ACM Trans. Inf. Syst. **22**(1), 5–53 (2004)
10. Hurley, N., Zhang, M.: Novelty and diversity in Top-N recommendation - analysis and evaluation. ACM Trans. Internet Technol. **10**(4), 14 (2011)

11. Bennett, J., Lanning, S.: The netflix prize. In: Proceedings of KDD Cup and Workshop, California, USA, 12 August 2007. ACM, New York (2007)
12. Kim, H.N., Saddik, A.E., Jung, J.G.: Leveraging personal photos to inferring friendships in social network services. Expert Syst. Appl. **39**(8), 6955–6966 (2012)
13. Koren, Y.: Factorization meets the neighborhood: a multifaceted collaborative filtering model. In: Proceedings of SIGKDD 2008, Las Vegas, NV, USA, 24–27 August 2008, pp. 426–434. ACM, New York (2008)
14. Koren, Y., Bell, R., Volinsky, C.: Matrix factorization techniques for recommender systems. Computer **42**(8), 30–37 (2009)
15. Linden, G., Smith, B., York, J.: Amazon.com recommendations: item-to-item collaborative filtering. IEEE Internet Comput. **7**(1), 76–80 (2003)
16. Lucas, J.P., Luz, N., Moreno, M.N., Anacleto, R., Almeida Figueiredo, A., Martins, C.: A hybrid recommendation approach for a tourism system. Expert Syst. Appl. **40**(9), 3532–3550 (2013)
17. McNee, S.M., Riedl, J., Konstan, J.A.: Being accurate is not enough: how accuracy metrics have hurt recommender systems. In: Proceedings of CHI EA 2006, Montréal, Canada, 22–27 April 2006, pp. 1097–1101. ACM, New York (2006)
18. Murakami, T., Mori, K., Orihara, R.: Metrics for evaluating the serendipity of recommendation lists. In: Satoh, K., Inokuchi, A., Nagao, K., Kawamura, T. (eds.) JSAI 2007. LNCS (LNAI), vol. 4914, pp. 40–46. Springer, Heidelberg (2008)
19. Özbal, G., Karaman, H., Alpaslan, F.N.: A content-boosted collaborative filtering approach for movie recommendation based on local and global similarity and missing data prediction. Comp. J. **54**(9), 1535–1546 (2011)
20. Padmanabhan, B., Tuzhilin, A.: A belief-driven method for discovering unexpected patterns. In: Proceedings of SIGKDD 1998, New York, NY, USA, 27–31 August 1998, pp. 94–100. AAAI Press, California (1998)
21. Qiuxia, L., Chen, T., Zhang, W., Yang, D., Yu, Y.: Serendipitous personalized ranking for Top-N recommendation. In: Proceedings WI 2012, Macau, China, 4–7 December 2012, pp. 258–264. IEEE Computer Society, Washington (2012)
22. Resnick, P., Iacovou, N., Suchak, M., Bergstrom, P., Riedl, J.: Grouplens: an open architecture for collaborative filtering of Netnews. In: Proceedings of CSCW 1994, Chapel Hill, North Carolina, USA, 22–26 October 1994, pp. 175–186. ACM, New York (1994)
23. Sarwar, B., Karypis, G., Konstan, J., Riedl, J.: Item-based collaborative filtering recommendation algorithms. In: Proceedings of WWW 2001, Hong Kong, China, 1–5 May 2001, pp. 285–295. ACM, New York (2001)
24. Steck, H.: Training and testing of recommender systems on data missing not at random. In: Proceedings of SIGKDD 2010, Washington, DC, USA, 25–28 July 2010, pp. 713–722. ACM, New York (2010)
25. Ziegler, C.N., McNee, S.M., Konstan, J.A., Lausen, G.: Improving recommendation lists through topic diversification. In: Proceedings of WWW 2005, Chiba, Japan, 10–14 May 2005, pp. 22–32. ACM, New York (2005)

Data Mining in Environment

An Approach for Predicting River Water Quality Using Data Mining Technique

Bharat B. Gulyani[1(✉)], J. Alamelu Mangai[1], and Arshia Fathima[2]

[1] BITS Pilani, Dubai Campus, P.O. Box 345055
Dubai International Academic City, UAE
gulyanibb@gmail.com, {gulyani,mangai}@dubai.
bits-pilani.ac.in
[2] University of California, Berkeley, CA, USA
arshiafathima92@gmail.com

Abstract. Water contains many chemical, physical, and biological impurities. Some impurities are benign while others are toxic. The quality of water is defined in terms of its physical, chemical, and biological parameters and ascertaining its quality is crucial before use for various intended purposes such as potable water, agricultural, industrial, etc. Various water analysis methods are employed to determine water quality parameters such as DO, COD, BOD, pH, TDS, salinity, chlorophyll-a, coli form, and organic contaminants such as pesticides. The list of potential water contaminants is exhaustive and impractical to test for in its entirety. Such water testing is sometimes costly and time consuming. This paper attempts to present application of data mining technique to build a model to predict a widely used gross water quality parameter called Biochemical oxygen demand (BOD). BOD is a measure of the amount of dissolved oxygen used by microbial oxidation of organic matter in wastewater. The standard method for measuring BOD is a 5-day process. Dilution of sample, constant pH and nutrient content besides the temperature of 20 °C and dark area are required for correct results. High levels of nitrogen compounds yield false BOD results. Winkler titration which is also used to measure BOD is a chemical intensive process. Hence an automatic prediction model for BOD has been sought for accurate, cost-effective and time saving measurement. Based on data available for BOD measurements, this paper describes the development of a prediction model for BOD using a technique of data mining, namely, support vector machines (SVM). A correlation coefficient of 0.9471 and RMSE of 0.5019 was obtained for the BOD prediction model on river water quality data. The performance of the proposed model was also compared with two other models namely artificial neural network (ANN) and regression by discretization. Simulation results show that the proposed model performs better than the other two in terms of correlation coefficient and RMSE.

Keywords: Biochemical oxygen demand (BOD) · Data mining · Support vector machines · Multiple regression · Correlation coefficient

© Springer International Publishing Switzerland 2015
P. Perner (Ed.): ICDM 2015, LNAI 9165, pp. 233–243, 2015.
DOI: 10.1007/978-3-319-20910-4_17

1 Introduction

Water quality monitoring and control is critical for the survival of modern civilization because degraded water quality means that desired uses are not possible or not safe. Water quality parameters such as dissolved oxygen (DO), biochemical oxygen demand (BOD), and chemical oxygen demand (COD) are used to determine the pollution levels. The degradation of organic and inorganic matter in water is important for the health of the ecosystem. Biodegradable matter consists of organics such as starches, fats, proteins, and esters that can be utilized for food by the microbes naturally occurring in the water [1].

Microbial degradation of dissolved matter can be accompanied by oxidation or reduction. Although both may occur simultaneously, oxidation process is predominant when oxygen is available [1]. Thus, in liquid wastes, dissolved oxygen is the factor that determines whether biological changes are brought about by aerobic or anaerobic organisms. Rate of biochemical oxidation is determined by residual dissolved oxygen measures at regular intervals of time [2].

Biochemical oxygen demand (BOD) is defined as the amount of oxygen required by living organisms while stabilizing decomposable organic matter under aerobic conditions. The BOD5 test is bioassay procedure that involves measuring dissolved oxygen consumption by microbes utilizing the organic matter in water under similar conditions found in nature [2]. The standard test is performed over 5 days at 20 °C by incubating small sample of water in BOD bottle saturated with oxygen and containing microbial seed and nutrients. The BOD value is obtained by finding the difference in dissolved oxygen content at the start of experiment and after the incubation period [3].

BOD measurement is a time consuming and tedious test that requires caution at each step for accurate results.BOD test must be performed in an environment suitable for biological growth that includes adequate amounts of nutrients and oxygen and absence of toxic substances. The nitrifying bacteria responsible for oxidation of nitrogenous compounds are usually present in small amounts and also have a slow reproductive rate. Hence 6–10 days are required to exert measurable oxygen demand for these compounds. However if these bacteria are present in large amounts, interference caused by nitrification will lead to erroneous BOD results [3]. Besides providing adequate oxygen for microbial growth, care must be taken to protect the test sample from air to prevent re-aeration as dissolved oxygen levels diminish over the incubation period. Significant population of mixed organisms of soil origin must be used to simulate natural conditions [2]. Because of these problems and uncertainties involved, efficient and time-saving prediction models to estimate and predict BOD have widely been investigated and analyzed.

Data mining techniques help in recognizing hidden patterns and relations within a given data. Data mining techniques have a wide range of applications in medical diagnosis, target marketing, computer security etc. for classification and prediction purposes. Many algorithms such as Artificial Neural Network (ANN), Support Vector Machine (SVM), regression, etc. are used for prediction and forecasting. Techniques such as neural networks, decision trees are considered unstable while K-nearest neighbours, linear regression and logistic regression are considered stable [4]. This paper proposes SVM with Pearson function based universal kernel for predicting BOD in river water from a data mining perspective. The predictive capability of the model is also compared with two other models namely ANN and regression by discretization for the same data set.

2 Literature Review

2.1 Data Mining Techniques in Water Quality Applications

Application of ecological data-mining techniques to develop prediction models for BOD has been active research area since early 2000's. Most of the prediction models found in literature are based on Artificial Neural Networks (ANN) with various input parameters. ANN is useful in determining non-linear input-output relations among variables. It is computationally fast and requires few input parameters and input conditions. However, large training set covering maximum general cases is required for accurate prediction capability [5]. It is also difficult to predict for an event that has not occurred in the training data for ANN [6].

ANN was applied for prediction of BOD in Yuqiao reservoir, China with correlation coefficient of 0.8537 and average error of 2.56 % [7]. For training the ANN, back propagation and Levenberg-Marquardt (LM) algorithm were applied. Back propagation allows for reaching the extreme minimum values and is slow to converge while LM algorithm provides for low error and high speed computation. ANN models have also used cascade correlation algorithm for training a multi-layer feed forward ANN models for prediction of conductivity, dissolved oxygen and 4 other parameters [8]. For calculation of weights in ANN, Kalman's learning rule was applied by them. A recent technique that has been applied for pattern recognition and prediction for water quality samples was particle swarm optimization based ANN that had 80 % training sample accuracy with acceptable error percentage [9].

k-Nearest Neighbors (kNN), support vector machines (SVM) and self organizing maps (SOM) were applied for forecasting BOD and suspended solids using 8 input parameters including turbidity, conductivity, redox- potential, temperature, dissolved oxygen and pH. SOM gave the best prediction results for BOD [10]. Classification and regression trees (CART) and multi-layer perceptron (MLP) were used for classification of water quality canals in Bangkok, Thailand [11]. Few data mining techniques that have been applied for total suspended solids (TSS) include MLP, kNN, multi-variate adaptive regression spline, SVM and random forest algorithms [12]. From above mentioned techniques, ANN has been predominantly used and has given accurate predictions. Apart from ANN, SOM developed has also given a correlation of approx 0.99 [10].

2.2 Support Vector Machine (SVM) Theory

The Support Vector Machine (SVM) is a classification method that belongs to the category of kernel methods. In kernel methods the dependence on data is via dot products only. This gives the advantages of defining non-linear decision boundaries as well as using kernel functions to apply classifier to data with no fixed dimension vector space. SVMs can be applied to high-dimensional data not only for classification but also for prediction. A linear classifier is based on the linear discrimination function given by the following equation which defines a hyperplane. The hyperplane divides the space into two according to the function f(x) in the Eq. 1. The hyperplane equation is obtained by equating f(x) = 0 in the following Eq. (1):

$$f(x) = w^T x + b \qquad (1)$$

Where $w^T x = \sum_i w_i x_i$ defines the dot product between the weight vector (w) and the input values (x); "b" is the bias which translates the hyperplane from origin.

A classifier is linear if the decision boundary defined by the hyperplane is linear else it is non-linear classifier. When the input space is given by a non-linear function $\Phi(x)$, the obtained Eq. (2) gives us a non-linear classifier [13]:

$$f(x) = w^T \Phi(x) + b \qquad (2)$$

Support vector margins are determined by the support vectors which are the instances lying on the margin of the decision boundary. The SVM is divided into two categories: the hard margin SVM that does not allow misclassification and the soft margin SVM that allows misclassification [13].

Mathematical modeling of SVM with hard margin is given by the following quadratic programming formulation. Dual form is obtained by application of Lagrangian optimization for quadratic problems. These problems are then subjected to Karush-Kuhn-Tucker conditions to get a solution. For hard margin SVM to get the maximum margin classifier the following constrained optimization problem is applied:

$$\text{minimize} \frac{1}{2} \|w\|^2 \qquad (3)$$

Subject to

$$y_i(w^T.x_i + b) \geq 1 \text{ where } i = 1,2,\ldots\ldots,n \qquad (4)$$

Dual Form for hard margin SVM after application of Lagrangian optimization is:

$$\text{minimize} \frac{1}{2} \sum_{i=1}^{N} \sum_{j=1}^{N} y_i y_j (x_i.x_j) \alpha_i \alpha_j - \sum_{j=1}^{N} \alpha_j \qquad (5)$$

Subject to

$$\sum_{i=1}^{N} y_i \alpha_i = 0 \text{ where } \alpha \geq 0, \text{ for } i = 1, 2, \ldots n \qquad (6)$$

Soft margin SVM includes the inequality constraints with ξ_i which are also called as slack variables allowing misclassifications. The parameter C ($C > 0$) called the complexity parameter is added to the optimization problem. This parameter is a trade-off between maximizing margin and allowance for misclassifications. The dual form for soft margin SVM is given below. The optimization problem will have the penalty for misclassification included as shown in the following equations:

$$\text{minimize } \frac{1}{2}||w||^2 + C\sum_{i=1}^{N}\xi_i \tag{7}$$

Subject to

$$y_i(w.x_i + b) \geq 1 - \xi_i, \text{where } i = 1, 2, \ldots.n \text{ and } \xi_i \geq 0 \tag{8}$$

Using the Lagrange multipliers the dual form of Soft margin SVM is:

$$\text{maximize} \sum_{i=1}^{N}\alpha_i - \frac{1}{2}\sum_{i=1}^{N}\sum_{j=1}^{N} y_iy_j(x_i^T.x_j)\alpha_i\alpha_j \tag{9}$$

Subject to

$$\sum_{i=1}^{N} y_i\alpha_i = 0, \text{with } 0 \leq \alpha_i \leq C \tag{10}$$

For non-linear support vector machines the dot product $(x\_i.x\_j)$ is replaced by $\Phi(x\_i).\Phi(x\_j)$ by the Kernel function $K(xi.xj)$ [14]. The common kernel functions used in SVMs are as follows [15]:

Polynomial kernel:

$$K(x_i, x_j)=(x_i.x_j)^d \text{ Or } K(x_i, x_j)=((x_i.x_j) + 1)^d \text{ where d} = 1,\ldots\ldots.n \tag{11}$$

Radial Basis function:

$$K(x_i, x_j) = \exp\left\{-\frac{(x_i - x_j)^2}{\sigma^2}\right\} \tag{12}$$

Sequential Minimal Optimization (SMO) is an algorithm used to solve the SVM QP problem effectively and efficiently. SMO chooses to solve the smallest possible optimization problem at every step solving two Lagrange multipliers analytically which is an advantage. SMO also performs well for SVM with sparse or non-linear inputs as the kernel computation time is reduced thereby speeding the SMO calculations [16].

Though SVM provides a highly accurate model for classification/prediction, any changes in the support vectors result in changes in the model itself. This is the reason why large data sets are required as a general representation of the population for classification. SVM classifiers are also sensitive to scaling of the features which require normalization for accurate classification. The choice of kernel function and longer training time are some other disadvantages for SVM [14].

2.3   SVM Applications in Water Quality Prediction

For prediction of BOD for wetlands based on input variables - conductivity, turbidity, redox potential, dissolved oxygen, water temperature and pH, SVM was used and the coefficient of determination was about 0.71 [10]. A hybrid approach of SVM with particle swarm optimization (PSO) applied to predict water quality parameters for Heishui river, China performed better than back-propagation neural networks [17]. To predict effluent parameters of wastewater such as COD, BOD, Ammonium-Nitrogen and TSS, Least Square Support Vector Regression (LS-SVR) with PSO was found to give accurate results with minimum error [18]. For prediction the total suspended solids (TSS) in waste-water based on carbonaceous bio-chemical oxygen demand (CBOD) and influent rate, Neural networks were found to perform better than SVM [12]. For water quality prediction model of Johor river in Malaysia, SVM was found to out-perform ensemble neural networks with 5 % of error distribution for DO, BOD and COD prediction [19]. Using ANOVA as kernel function for SVM, the prediction of dichotomized value of DO levels in Chini and Bera lake (Malaysia) was found to have an accuracy of 74 % [20].

3   Performance Evaluation Using Statistical Metrics

In this study, root mean squared error (RMSE) and correlation coefficient was used for analyzing the performance of the model.

The RMSE is given by [21]:

$$RMSE = \sqrt{\sum_{i=1}^{N} \frac{(p_i - a_i)^2}{N}} \qquad (13)$$

Where pi is the predicted value for i^{th} instance, a_i is the actual value for i^{th} instance and N is the total number of instances.

Correlation coefficient (also called Pearson correlation coefficient) measures the degree of linear relation for two variables. For a correlation coefficient of 1, the variables are perfectly correlated while correlation coefficient of zero implies no correlation exists between the two variables. Negative correlation coefficients indicate inverse relation between the variables. Larger values for correlation coefficient indicate better performance of the model. Correlation coefficient between actual and predicted values is defined as [22]

$$Correlation\ coefficient = \frac{S_{pa}}{\sqrt{S_p S_a}} \qquad (14)$$

Where

$$S_{pa} = \sum_i^N (p_i - \bar{p})(a_i - \bar{a}) \Big/ (N - 1) \qquad (15)$$

$$S_p = \sum_i^N (p_i - \bar{p})^2 \Big/ (N - 1) \qquad (16)$$

$$S_p = \sum_i^N (p_i - \bar{p})^2 \Big/ (N - 1) \qquad (17)$$

and $\bar{p}, \bar{a}$ are the averages respectively in the above equations.

Coefficient of performance, given by square of correlation coefficient, is used for analyzing the performance of the model [21]. For a classifier the total expected error analysis is given by bias-variance decomposition. The total expected error is the sum of bias and variance. For numeric prediction, bias is equal to mean square error expected when averaged over models built from all possible training datasets of same size.

4 Results and Discussion

The data set were taken from Department of Environment, Food and Rural Affairs, UK Government [23]. For this paper a large data set from North-east region was considered with data available from 1980–2010. Parameters include temperature (°C), pH, conductivity (μS/cm), suspended solids (mg/L), DO/dissolved oxygen (mg/L), ammoniacal nitrogen (mg/L), nitrate (mg/L),nitrate (mg/L), chloride (mg/L), total alkalinity (mg/L), orthophosphate (mg/L) and BOD (mg/L). These attributes can be easily measured with the help of probe in case of DO, temperature and conductivity, gravimetry for suspended solids and standard titration techniques using common chemicals for other parameters. These parameters were chosen based on their observed effects on BOD measurement. The amount of nitrogen based compounds have shown to affect BOD measurements [21]. The temperature influences the solubility of dissolved gases in water while pH is an indicator of acidic or alkaline substances while conductivity measures the amount of ions present that affect microbial growth thereby indirectly influencing the BOD values.

4.1 Data Analysis

The Pearson Correlation coefficient was calculated for the parameters using Excel Data Analysis tool. The results are shown in Table 1 below. The correlation matrix verifies the association between the parameters chosen for building the model and output parameter of the model. Higher correlation has been found between nitrate, ammoniacal nitrogen and conductivity. Effects of nitrification due to nitrates and ammonia on BOD measurement has been explained in Sect. 1. Dissolved oxygen has a negative correlation with BOD indicating that a decrease in dissolved oxygen of river will result in increased demand for microbial oxidation (BOD). Temperature of river has the least correlation with BOD as the temperature for standard BOD test has to be kept constant at 20 °C.

Table 1. Associated strength of input parameters with BOD using Pearson correlation coefficient

Parameter	Temp	pH	Cond	SS	DO	Amm	Nitrite	Nitrate	Chloride	Alkaline	Orthop	BOD
Temp	1.000											
pH	-0.216	1.000										
Cond	0.327	-0.498	1.000									
SS	0.115	-0.397	0.472	1.000								
DO	-0.668	0.468	-0.563	-0.352	1.000							
Amm	0.139	-0.444	0.770	0.536	-0.404	1.000						
Nitrite	0.360	-0.621	0.874	0.520	-0.598	0.823	1.000					
Nitrate	0.359	-0.362	0.806	0.287	-0.567	0.319	0.615	1.000				
Chloride	0.305	-0.517	0.975	0.503	-0.538	0.848	0.898	0.729	1.000			
Alkaline	0.253	-0.291	0.848	0.300	-0.496	0.636	0.691	0.769	0.816	1.000		
Orthop	0.374	-0.386	0.880	0.343	-0.570	0.582	0.778	0.849	0.835	0.791	1.00	
BOD	0.297	-0.660	0.783	0.579	-0.637	0.777	0.890	0.533	0.804	0.585	0.71	1.0

4.2 Proposed Model Results and Analysis

The model proposed in this paper for predicting BOD is also compared with some of the other existing models such as artificial neural network (ANN) and regression by discretization. Many induction tasks are found to benefit from discrete attributes than numeric attributes. For regression by discretization the numeric attributes in the input are discretized into 10 number of intervals. The base classifier used is the decision tree based classifier. The models were induced using 10 fold cross validation in WEKA [24]. Table 2 compares the correlation coefficient and RMSE of the SVM model with different kernel functions.

Table 2. Performance of SVM with various kernel functions

Kernel function	Correlation coefficient	RMSE
Poly-Kernel	0.9109	0.6506
Normalized Poly-Kernel	0.9421	0.5259
PUK (Pearson VII function based universal kernel)	0.9471	0.5019
RBF kernel	0.8936	0.802

It can be inferred from the results in Table 2, that SVM based prediction with Pearson function based universal kernel has the highest correlation coefficient and least

RMSE. The performance of the proposed model is also compared with ANN and regression with discrete features. Table 3 shows the parameter settings used to induce these models for comparative analysis.

Table 3. Parameter settings for model induction

Model	Parameter settings
ANN	Hidden layers = (attributes + classes)/2 = 6;
	Training time = 500 epochs;
	Learning rate = 0.3; Momentum = 0.2;
	Normalize attributes & numeric class;
	Validation threshold = 20
Regression by discretization	Classifier = J48;
	Numbins = 10;
Proposed model (for BOD prediction) - SMO regression	Complexity parameter = 1.0;
	Filter type = Normalize attributes;
	Kernel = PUK (Pearson VII function based universal kernel);
	RegOptimtizer- RegSMOImproved

The two performance metrics namely correlation coefficient and RMSE of all three models on the test data set are given in Table 4.

Table 4. Comparative performance analysis of BOD prediction models

Model	Correlation coefficient	RMSE
ANN	0.9168	0.6728
Regression by discretization	0.933	0.4256
Proposed model for BOD prediction (SMO regression)	0.9471	0.5019

Results in Table 4 show that the proposed model for predicting BOD using SVM with Pearson function based kernel has a high correlation coefficient when compared with the other two models. Regression using discrete attributes also performs better than ANN and has the least RMSE among all three models. The ANN model has the least correlation coefficient and a high RMSE among all three models.

5 Conclusions

This paper has proposed an approach using data mining techniques for predicting BOD in river water. The proposed model uses support vector machine (SVM) based regression technique for prediction. Pearson function based kernel is used to train SVM. The predictive capability of the model is evaluated using two metrics namely correlation coefficient and RMSE. A comparative analysis of the proposed method is also done with two other widely used methods namely artificial neural network (ANN) and regression by discretization. Simulation results prove that the proposed model has the highest correlation coefficient among all three methods.

References

1. Peavy, H.S., Donald, R., Tchobanoglous, G.: Environmental Engineering. Mc-Graw Hill, London (1985). International Edition
2. Sawyer, C.N., McCarthy, L., et al.: Chemistry for Environmental Engineering and Science, 5th edn. Tata Mc-Graw Hill, New Delhi (2003)
3. Metcalf, Eddy: Wastewater Engineering-Treatment and Reuse, 4th edn. Mc-Graw Hill, New Delhi (2003)
4. Zheng, Z., Padmanabhan, B.: Constructing ensembles from data envelopment analysis. INFORMS J. Comput. **19**(4), 486–496 (2007)
5. Najah, A., Elshafie, A., Karim, A., Jaffar, O.: Prediction of Johor river water quality parameters using artificial neural networks. Eur. J. Sci. Res. **28**(3), 422–435 (2009)
6. Talib, A., Abu Hassan, Y., Abdul Rahman, N.N.: Predicting biochemical oxygen demand as indicator of river pollution using artificial neural networks. In: 18th World IMACS/MODSIM Congress, Cairns, Australia (2009)
7. Ying, Z., Jun, N., Fu-yi, C., Liang, G.: Water quality forecast through application of BP neural network at Yuqiao reservoir. J. Zhejiang Univ. **8**(9), 1482–1487 (2007)
8. Diamantopoulou, M.J., et al.: Use of neural network technique for prediction of water quality parameters of Axios river in Northern Greece. Eur. Water **11**(12), 55–62 (2005)
9. Zhou, C., Gao, L., Gao, H., Peng, C.: Pattern classification and prediction of water quality by neural network with particle swarm optimization. In: Proceedings of the 6th World Congress on Intelligent Control and Automation, Dalian, China (2006)
10. Lee, B.-H., Scholz, M.: A comparative study: prediction of constructed treatment wetland performance with k-nearest neighbors and neural networks. Water Air Soil Pollut. **174**(1–4), 279–301 (2006)
11. Sirilak, A., Siripun, S.: Classification and regression trees and MLP neural network to classify water quality of canals in Bangkok, Thailand. Int. J. Intell. Comput. Res. **1**(1/2) (2010)
12. Verma, A., Wei, X., Kusiak, A.: Predicting the total suspended solids in wastewater: a datamining approach. Eng. Appl. Artif. Intell. **26**(4), 1366–1372 (2013)
13. Ben-hur, A., Weston, J.: A user's guide to support vector machines. In: Carugo, O., Eisenhaber, F. (eds.) Data Mining Techniques for Life Sciences, vol. 609, pp. 223–239. Humana Press, Totowa (2010)
14. Hongyu, S.: Multiple Kernel Learning with Sequential Minimal Optimization (SMO) Algorithm. Aalto University School of Science, Espoo (2012)
15. Stitson, M.O., Weston, J.A.E., Gammerman, A., Vovk, V., Vapnik, V.: Theory of support vector machines. Technical report – CSD-TR-96-17 (1996)
16. Platt, J.C.: A fast algorithm for training support vector machines, Microsoft Research. Technical report (1998)
17. Xuan, W., Jiak, L.V., Deti, X.: A hybrid approach of support vector machine with particle swarm optimization for water quality prediction. In: 5th International Conference on Computer Science & Education, Hefei, China (2010)
18. Huang, Z., Luo, J., Li, X., Zhou, Y.: Prediction of effluent parameters of wastewater treatment plant based on improved least square support vector machine with PSO. In: 1st International Conference on Information Science & Engineering – ICISE, Nanjing, China (2009)
19. Najah, A., El-Shafie, A., Karim, O.A., Jaafar, O., El-Shafie, A.H.: An application of different artificial intelligences techniques for water quality prediction. Int. J. Phys. Sci. **6**(22), 5298–5308 (2011)

20. Malek, S., Mosleh, M., Sharifah, M.S.: Dissolved oxygen prediction using support vector machines. Int. Comput. Inf. Sci. Eng. World Acad. Sci. Eng. Technol. **8**(1), 46–50 (2014)
21. Singh, K.P., Basant, A., Malik, A., Jain, G.: Artificial neural network modeling of the river water quality—a case study. Ecol. Model. **220**, 888–895 (2009)
22. Witten, I.H., Frank, E.: Data Mining-Practical Machine Learning Tools and Technology with Java Implementations. Morgan Kauffman Publications, California (2000)
23. http://data.gov.uk/dataset/river-water-quality-regions
24. Hall, M., Frank, E., Holmes, G., Pfahringer, B., Reutemann, P., Witten, I.H.: The WEKA data mining software: an update. SIGKDD Explor. **11**(1), 1010–1012 (2009)

Adaptive Learning

An Efficient Data Mining Approach to Concept Map Generation for Adaptive Learning

Xiaopeng Huang[1(✉)], Kyeong Yang[2], and Victor B. Lawrence[2]

[1] SmileK12 Inc., Freehold, NJ, USA
xhuang@smilek12.com
[2] Stevens Institute of Technology, Hoboken, NJ, USA
{kyeong.yang, victor.lawrence}@stevens.edu

Abstract. Data mining has recently drawn an increasing interest as an effective approach to generation of a concept map in an adaptive learning platform that provides students with personalized learning guidance. Although it has seen significant progresses, the data mining-based concept map generation needs to be further improved both in complexity and accuracy for wide acceptance in actual education services. This paper aims to improve the accuracy of concept map by considering both wrong-to-wrong and correct-to-correct relationships of questions, and by adopting more accurate formulas in calculation of relevance degrees between concepts. Through simulations using a set of concepts, questions, and student test records sampled from a practical courseware, we show that the proposed approach can generate a more accurate and robust concept map at an acceptable additional complexity.

Keywords: Data mining · Concept map · Adaptive learning · Apriori algorithm · Question-to-concept relationship · Two paths algorithm

1 Introduction

With recent advancement in data mining-based adaptive learning techniques [1, 6, 9], there have been proposed several adaptive learning systems that offer students customized course content in accordance with their aptitudes and learning results [5, 10, 18]. Adaptive learning usually involves adaptive content aggregation, adaptive recommendation, adaptive presentation, adaptive navigation (concept map), students' knowledge score estimation, and adaptive collaboration. Adaptive learning also builds a model for student characteristics and uses that model in interacting with the student to provide the best possible learning experience [4].

A concept map, consisting of key concepts and links that illustrate relationships among those concepts, plays a critical role in providing adaptive learning guidance to students as well as in evaluating students' knowledge level of various concepts. However, it is difficult and time consuming to manually create a concept map for a course or a subject. Thus, there have been efforts to automatically create concept maps based on students' historical test records.

The concept map stems from the need to show explicitly how new concepts and propositions are integrated into the learner's cognitive structure. Novak proposed the

© Springer International Publishing Switzerland 2015
P. Perner (Ed.): ICDM 2015, LNAI 9165, pp. 247–260, 2015.
DOI: 10.1007/978-3-319-20910-4_18

concept map in 1998 to organize and represent the knowledge as a network that consists of nodes as concepts and links as the relations among concepts [11]. Thereafter, in 2001 Tsai et al. proposed a learning algorithm, integrating data mining, fuzzy set theory, and prediction mechanism, to help teachers analyze, refine and reorganize the teaching and assessment materials based on students' learning records [15]. In 2007, Tseng et al. proposed a two-phase concept map construction (TP-CMC) approach to automatically construct the concept map by students' historical test records [16]. In 2008, Bai and Chen developed an automatic concept map generation algorithm that applies fuzzy rules on students' test records for an adaptive learning system [2]. They further applied the concept map to evaluation of the learning achievement of students [3]. In 2011, Sarem et al. also employed the fuzzy set theory in the concept map generation to take into account numerical test scores rather than the binary score for each question [13]. Lee et al. developed an intelligent concept diagnostic system applying the Apriori algorithm to concept map generation to enable teachers to instantly diagnose the learning barriers and misconception of learners [8].

However, these papers do not include a fair evaluation of the accuracy and analysis of the complexity. Only a handful of papers tried practical data in evaluation of the accuracy [18], and most papers employed hypothetical questions and students in data mining for concept map generation, making conclusions of the papers less convincible. In addition, most papers for concept map generation do not employ a concept map evaluation process. Therefore, it is hard to prove the accuracy and reliability of the generated concept maps. In [18], Yang et al. have evaluated their algorithm, but it was performed manually by teachers since there is no standard way to automatically evaluate concept maps. We have not found education materials that are commonly used as test data by the community of data mining researchers.

The complexity of a concept map generation algorithm would dramatically increase as the number of test records and concepts increases, and can become a critical issue that would prevent such concept map generation from being adopted in adaptive learning platforms. The classification-based approach is an effort to significantly reduce the complexity of concept map generation by performing the data mining process on subsets of test records that are usually much smaller than the original records [7]. However, applying data mining on the reduced size of subset data could degrade the accuracy and robustness of the generated concept map. A similar statement could be made for cases where the data mining process is applied to a small set of data.

This paper aims to improve the accuracy of concept map by considering both wrong-to-wrong and correct-to-correct relationships of questions, especially when the data set is small for any reason. This paper also proposes accurate formulas for calculation of relevance degrees between concepts to further improve the accuracy.

The rest of this paper is organized as follows: Sect. 2 describes the proposed algorithm. Section 3 presents an example of generating a concept map using the proposed algorithm and discusses the result. Section 4 concludes the paper and discusses future works.

2 Algorithm Description

2.1 Overview of the Proposed Algorithm

The proposed data mining approach for concept map generation is depicted in Fig. 1, where the same data mining process is applied to both the wrong-to-wrong relationship (Path 1) and correct-to-correct relationship (Path 2) of questions and results from the two paths are merged to calculate resultant relevance degrees that will form a concept map together.

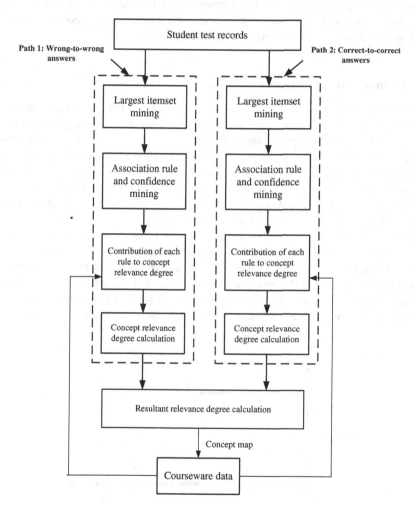

Fig. 1. Diagram of the proposed data mining approach for concept map generation

This algorithm involves two types of input data: student test records (dynamic data) and question-to-concept relevance scores (deterministic data) in courseware, and adopts an association rule algorithm (e.g., Apriori algorithm) [17, 19] to perform data mining for concept map generation. It first utilizes the largest itemset mining to reduce the number of rules with a little degradation in accuracy, and then conducts association rule and confidence mining to select association rules to be considered in calculation of relevance degrees among concepts and calculates their confidence (probability) values. For each of the survived association rules, we calculate its contribution to the relevance degree of any two concepts by using the respective confidence value and question-to-concept relevance data obtained from the courseware. This step connects the courseware data and student test records. Then, we calculate the relevance degree of two concepts by applying a formula on the contributions of all association rules to the respective concept relevance. The same process is performed on the correct-to-correct relationship. Finally, the process combines the two relevance degrees obtained from the two paths, and outputs the concept relevance degrees.

2.2   Definition of Test Records

The proposed algorithm processes two types of data for concept map generation: one is student test record (e.g., student test result for questions and student grades), and the other is question-to-concept relationship which may be generated by experienced teachers and obtained from courseware database. Student test record is practical and dynamic data, which is processed by data mining algorithms, such as fuzzy clustering algorithm, classification algorithm, and association rule algorithm [12, 14]. In order to get accurate results of concept map generation, we pre-process student test record and adopt a proper function model to deal with test record.

For a common function model, the grade range of any question is $[0, R]$, where R is an integral number larger than 0. The grade can also be classified into N clusters so as to make the data mining process simpler and/or achieve a more accurate data mining result. Figure 2 is the function model we use in this paper ($R = 1$ and $N = 2$). This function only has two classification results of 0 and 1, where 0 represents that a student got the correct answer, and 1 represents that the student got the wrong answer.

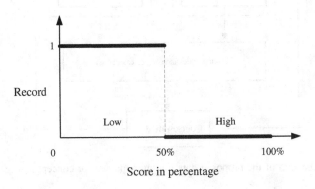

Fig. 2.   Example function model of student test records

We also define matrixes for student test results (grade for each question) and relationships of questions to various concepts. Assume there are N_S students (S_1, S_2, $\cdots$, S_{Ns}) and N_Q questions (Q_1, Q_2, $\cdots$, Q_{NQ}) in test record. Then, the matrix for student test results is defined as in Eq. (1):

$$
M_{QS} = \begin{array}{c} \\ S_1 \\ S_2 \\ \vdots \\ S_{Ns} \end{array} \begin{array}{cccc} Q_1 & Q_2 & \cdots & Q_{N_Q} \\ \left[\begin{array}{cccc} qs_{11} & qs_{12} & \cdots & qs_{1N_Q} \\ qs_{21} & qs_{22} & \cdots & qs_{2N_Q} \\ \vdots & \vdots & \vdots & \vdots \\ qs_{N_s1} & qs_{N_s2} & \cdots & qs_{N_sN_Q} \end{array} \right] \end{array}
\tag{1}
$$

where qs_{ij} denotes the grade of the ith student S_i for the jth question Q_j, and $qs_{ij} \in [0, R]$, $1 \leq i \leq N_S$, and $1 \leq j \leq N_Q$.

For the case of N_C concepts (C_1, C_2, $\cdots$, C_{Nc}) involved in the tests, we define a question-to-concept relationship matrix as in Eq. (2):

$$
M_{QC} = \begin{array}{c} \\ Q_1 \\ Q_2 \\ \vdots \\ Q_{N_Q} \end{array} \begin{array}{cccc} C_1 & C_2 & \cdots & C_{N_C} \\ \left[\begin{array}{cccc} qc_{11} & qc_{12} & \cdots & qc_{1N_C} \\ qc_{21} & qc_{22} & \cdots & qc_{2N_C} \\ \vdots & \vdots & \vdots & \vdots \\ qc_{N_Q1} & qc_{N_Q2} & \cdots & qc_{N_QN_C} \end{array} \right] \end{array}
\tag{2}
$$

where qc_{ij} denotes the relevance score of the ith question Q_i to the jth concept C_j, $qc_{ij} \in [0, S]$, $1 \leq i \leq N_Q$, and $1 \leq j \leq N_C$. S denotes the range of relevance scores between questions and concepts. "$qc_{ij} = 0$" means that the ith question Q_i does not have any relation with the jth concept C_j.

2.3   Algorithm Description

The proposed algorithm is run with the student test record and relevance scores of each question to concepts that are stored in the courseware. The data mining process is performed independently for Path 1 and Path 2 using the same function, so we use Path 1 as an example to elaborate Steps 1-5 of this algorithm. The same steps are applied to Path 2 simultaneously.

The proposed algorithm is presented as follows:

Step 1: Perform the largest itemset mining for the wrong-to-wrong relationship to find the most frequent itemset by using an association rule algorithm, such as the Apriori algorithm. Setting the minimum support plays a critical role in the largest itemset mining process. Such itemsets whose support values are less than the minimum support are excluded in each step. In order to make the largest itemset mining process more effective, we set the minimum support, denoted as sup_{min}, based on the support values of the 1-itemsets, as in Eq. (3):

$$sup_{min} = \left[\frac{\sum sup(Q_i)}{N_Q}\right] - 1 \qquad (3)$$

where $sup(Q_i)$ is the support value of Q_i, and $[x]$ is the largest integer that doesn't exceed x.

Step 2: Based on the itemsets survived in the largest itemset mining process in Step 1, perform the association rule and confidence mining to generate association rules and their confidence values for the wrong-to-wrong relationship. Confidence is the probability of the event that a student also incorrectly answers the question Q_j when the student answers the question Q_i incorrectly. Therefore, the "$conf^{w}(Q_i \rightarrow Q_j)$" of an association rule "$Q_i \rightarrow Q_j$" is calculated as in Eq. (4):

$$conf^{w}(Q_i \rightarrow Q_j) = \frac{sup^{w}(Q_i, Q_j)}{sup^{w}(Q_i)} \qquad (4)$$

where $sup^{w}(Q_i, Q_j)$ denotes the support of the 2-itemset (Q_i, Q_j) of the wrong-to-wrong relationship and $sup^{w}(Q_i)$ denotes the support of the 1-itemset Q_i of wrong answers, $1 \le i \le N_Q$, $1 \le j \le N_Q$, and $i \ne j$.

Step 3: Using the relevance of multiple questions to various concepts provided from the courseware data in the format of the question-to-concept relationship matrix M_{QC} in Eq. (2), we calculate contribution of each association rule "$Q_i \rightarrow Q_j$" to the relevance degree of $C_x \rightarrow C_y$, denoted as $R_{cxcy}^{w}(Q_i \rightarrow Q_j)$, as in Eq. (5):

$$R_{c_xc_y}^{w}(Q_i \rightarrow Q_j) = conf^{w}(Q_i \rightarrow Q_j) \times \left(R_{Q_iC_x} \cdot R_{Q_jC_y} \cdot R_{Q_iC_y}\right)^{\frac{1}{3}} \qquad (5)$$

where $R_{Q_iC_x}$, $R_{Q_jC_y}$, and $R_{Q_iC_y}$ are the relevance scores of $Q_i \rightarrow C_x$, $Q_j \rightarrow C_y$, and $Q_i \rightarrow C_y$, respectively, for the wrong-to-wrong relationship that are provided by the courseware data.

Step 4: Calculate the relevance degree of $C_x \rightarrow C_y$ for the wrong-to-wrong relationship, R_{cxcy}^{w}, as the arithmetic average of relevance degrees of all (N_w) association rules:

$$R_{c_xc_y}^{w} = \sum_{r=1}^{N_w}(d_r \cdot d_r)/\sum_{r=1}^{N_w}d_r \qquad (6)$$

where $d_r = R_{cxcy}^{w}(Q_i \rightarrow Q_j)$ denotes the relevance degree for the rth association rule for the wrong- to-wrong relationship.

Step 5: Merge the relevance degrees of $C_x \rightarrow C_y$ for wrong-to-wrong and correct-to-correct relationships of questions to obtain the final relevance degree between two concepts, using Eq. (7):

$$R_{c_xc_y} = \alpha R_{c_xc_y}^{w} + (1 - \alpha)R_{c_xc_y}^{c} \qquad (7)$$

where α and $(1 - \alpha)$ are weights for the concept relevance degree for the wrong-to-wrong and the correct-to-correct relationships, respectively, and R^c_{cxcy} is the relevance degree of $C_x \rightarrow C_y$ for the correct-to-correct relationship of questions.

3 Simulations and Discussions

In this section, we present an example to elaborate the concept map generation process based on the algorithm described in Sect. 2.3.

3.1 Test Data for Student Records and Courseware

The test in our example[1] involves 4 questions, 4 concepts, and 5 students, and the student test record complies with the function model shown in Fig. 2, which includes only two binary values indicating correct (0) or wrong (1). Then, the input data matrixes in Eqs. (1) and (2) can be expressed respectively as in Eqs. (8) and (9):

$$
M_{QS} = \begin{array}{c} \\ S_1 \\ S_2 \\ S_3 \\ S_4 \\ S_5 \end{array}
\begin{array}{cccc} Q_1 & Q_2 & Q_3 & Q_4 \\ \left[\begin{array}{cccc} 0 & 1 & 0 & 1 \\ 0 & 1 & 1 & 1 \\ 1 & 0 & 1 & 1 \\ 0 & 0 & 0 & 1 \\ 0 & 0 & 1 & 1 \end{array}\right] \end{array} \tag{8}
$$

$$
M_{QC} = \begin{array}{c} \\ Q_1 \\ Q_2 \\ Q_3 \\ Q_4 \end{array}
\begin{array}{cccc} C_1 & C_2 & C_3 & C_4 \\ \left[\begin{array}{cccc} 8 & 2 & 0 & 0 \\ 2 & 6 & 1 & 1 \\ 0 & 3 & 6 & 1 \\ 0 & 1 & 2 & 7 \end{array}\right] \end{array} \tag{9}
$$

where in the matrix M_{QS}, 1 represents the wrong answer and 0 represents the correct answer; in the matrix M_{QC}, [1, 10] is the range of relevance score of each question to concept, and 0 between Q_i and C_j means that Q_i does not have any relation with C_j.

3.2 Concept Map Generation

This sub-section shows how to generate a concept map following the algorithm proposed in Sect. 2.3. Detailed steps are described as follows:

---

[1] The test data was generated by several students and questions in Liner Algebra & Geometry, and has been modified to make it suitable for the purpose of the paper. The difficulty level of concepts increases from C_1 to C_4.

Table 1. Result of 1-itemset mining based on the Apriori algorithm

Candidate (wrong-to-wrong)	Support (wrong-to-wrong)	Candidate (correct-to-correct)	Support (correct-to-correct)
Q_1	1	Q_1	4
Q_2	2	Q_2	3
Q_3	3	Q_3	2
Q_4	5	Q_4	0

Table 2. Result of 2-itemset mining based on the Apriori algorithm

Candidate (wrong-to-wrong)	Support (wrong-to-wrong)	Candidate (correct-to-correct)	Support (correct-to-correct)
$\{Q_1, Q_2\}$	0	$\{Q_1, Q_2\}$	2
$\{Q_1, Q_3\}$	1	$\{Q_1, Q_3\}$	2
$\{Q_1, Q_4\}$	1	$\{Q_2, Q_3\}$	1
$\{Q_2, Q_3\}$	1		
$\{Q_2, Q_4\}$	2		
$\{Q_3, Q_4\}$	1		

Table 3. Result of 3-itemset mining based on the Apriori algorithm

Candidate (wrong-to-wrong)	Support (wrong-to-wrong)	Candidate (correct-to-correct)	Support (correct-to-correct)
$\{Q_1, Q_3, Q_4\}$	1	$\{Q_1, Q_2, Q_3\}$	1
$\{Q_2, Q_3, Q_4\}$	1		

Table 4. Result of 4-itemset mining based on the Apriori algorithm

Candidate (wrong-to-wrong)	Support (wrong-to-wrong)	Candidate (correct-to-correct)	Support (correct-to-correct)
$\{Q_1, Q_2, Q_3, Q_4\}$	0	$\{Q_1, Q_2, Q_3, Q_4\}$	0

Step 1: Tables 1, 2, 3 and 4 show the largest itemset mining results of both wrong-to-wrong and correct-to-correct relationships of questions. The minimum support values for the wrong-to-wrong and correct-to-correct relationships are calculated using Eq. (3):

$$sup_{min}^{w} = [(1 + 2 + 3 + 5)/4] - 1 = 2 - 1 = 1 (\text{for wrong-to-wrong}) \qquad (10a)$$

$$sup_{min}^{c} = [(4 + 3 + 2 + 0)/4] - 1 = 2 - 1 = 1 (\text{for correct-to-correct}) \qquad (10b)$$

The minimum support equals to 1, indicating only itemsets having a support value larger than or equal to 1 are selected for further considerations. Tables 1, 2, 3 and 4 also show that the largest itemset is 3-itemset for both wrong-to-wrong and

Table 5. List of survived association rules and their confidences

Rule (wrong-to-wrong)	Confidence (wrong-to-wrong)	Rule (correct-to-correct)	Confidence (correct-to-correct)
Q_1 to Q_3	1	Q_1 to Q_2	0.5
Q_1 to Q_4	1	Q_1 to Q_3	0.5
Q_2 to Q_3	0.5	Q_2 to Q_1	0.67
Q_2 to Q_4	1	Q_2 to Q_3	0.33
Q_3 to Q_1	0.33	Q_3 to Q_1	1
Q_3 to Q_2	0.33	Q_3 to Q_2	0.5
Q_3 to Q_4	1		
Q_4 to Q_1	0.2		
Q_4 to Q_2	0.4		
Q_4 to Q_3	0.6		

correct-to-correct relationships. For more information about this step including the Apriori algorithm, please refer to [17] and [19].

Step 2: Based on the largest itemset results obtained in Step 1, we construct and select association rules and calculate their confidence values. Table 5 shows the survived rules and their confidence values for both wrong-to-wrong and correct-to-correct relationships, where the confidence values of "Rule Q_2 to Q_3" of the wrong-to-wrong and correct-to-correct relationships, as an example, are calculated as in Eqs. (11a) and (11b) by using Eq. (4):

$$conf^w(Q_2 \rightarrow Q_3) = sup^w(Q_2, Q_3)/sup^w(Q_2) = 1/2 = 0.5 \qquad (11a)$$

$$conf^c(Q_2 \rightarrow Q_3) = sup^c(Q_2, Q_3)/sup^c(Q_2) = 1/3 = 0.33 \qquad (11b)$$

Step 3: Calculate contribution of each association rule "$Q_i \rightarrow Q_j$" to the relevance degree of "$C_x \rightarrow C_y$" for both wrong-to-wrong and correct-to-correct relationships. Results are shown in Table 6 and Table 7. Equations (12a) and (12b) demonstrate how to calculate contribution of each association rule to the relevance degree by calculating contributions of "Rule Q_2 to Q_3" to the relevance degree of "C_1 to C_2" for wrong-to-wrong and correct-to-correct relationships, respectively:

$$R^w_{c_1c_2}(Q_2 \rightarrow Q_3) = 0.50 \times (2 \times 3 \times 6)^{1/3} = 1.65 \qquad (12a)$$

$$R^c_{c_1c_2}(Q_2 \rightarrow Q_3) = 0.33 \times (2 \times 3 \times 6)^{1/3} = 1.09 \qquad (12b)$$

Step 4: Calculate the concept relevance degree of $C_x \rightarrow C_y$ for both wrong-to-wrong and correct-to-correct relationships. Results are shown in Table 8 where the 2nd and 6th columns are for the wrong-to-wrong relationship (Path 1) and the 3rd and 7th columns are for the correct-to-correct relationship (Path 2). Equations (13a) and (13b) demonstrate how to calculate the concept relevance degree by calculating the relevance degree of "$C_1 \rightarrow C_2$" for wrong-to-wrong and correct-to-correct relationships, respectively:

Table 6. Contribution of each association rule "$Q_i \to Q_j$" to the relevance degree of "$C_x \to C_y$" (wrong-to-wrong relationship)

Rule	C_1 to C_2	Rule	C_1 to C_3	Rule	C_1 to C_4	Rule	C_2 to C_1
Q_1 to Q_3	3.63	Q_1 to Q_3	0.00	Q_1 to Q_3	0.00	Q_1 to Q_3	0.00
Q_1 to Q_4	2.52	Q_1 to Q_4	0.00	Q_1 to Q_4	0.00	Q_1 to Q_4	0.00
Q_2 to Q_3	1.65	Q_2 to Q_3	1.14	Q_2 to Q_3	0.63	Q_2 to Q_3	0.00
Q_2 to Q_4	2.29	Q_2 to Q_4	1.59	Q_2 to Q_4	2.41	Q_2 to Q_4	0.00
Q_3 to Q_1	0.00	Q_3 to Q_1	0.00	Q_3 to Q_1	0.00	Q_3 to Q_1	0.00
Q_3 to Q_2	0.00	Q_3 to Q_2	0.00	Q_3 to Q_2	0.00	Q_3 to Q_2	0.00
Q_3 to Q_4	0.00	Q_3 to Q_4	0.00	Q_3 to Q_4	0.00	Q_3 to Q_4	0.00
Q_4 to Q_1	0.00	Q_4 to Q_1	0.00	Q_4 to Q_1	0.00	Q_4 to Q_1	0.00
Q_4 to Q_2	0.00	Q_4 to Q_2	0.00	Q_4 to Q_2	0.00	Q_4 to Q_2	0.00
Q_4 to Q_3	0.00	Q_4 to Q_3	0.00	Q_4 to Q_3	0.00	Q_4 to Q_3	0.00

Rule	C_2 to C_3	Rule	C_2 to C_4	Rule	C_3 to C_1	Rule	C_3 to C_2
Q_1 to Q_3	0.00	Q_1 to Q_3	0.00	Q_1 to Q_3	0.00	Q_1 to Q_3	0.00
Q_1 to Q_4	0.00	Q_1 to Q_4	0.00	Q_1 to Q_4	0.00	Q_1 to Q_4	0.00
Q_2 to Q_3	1.65	Q_2 to Q_3	0.91	Q_2 to Q_3	0.00	Q_2 to Q_3	1.31
Q_2 to Q_4	2.29	Q_2 to Q_4	3.48	Q_2 to Q_4	0.00	Q_2 to Q_4	1.82
Q_3 to Q_1	0.00	Q_3 to Q_1	0.00	Q_3 to Q_1	0.00	Q_3 to Q_1	1.09
Q_3 to Q_2	0.86	Q_3 to Q_2	0.48	Q_3 to Q_2	0.00	Q_3 to Q_2	1.57
Q_3 to Q_4	3.30	Q_3 to Q_4	2.76	Q_3 to Q_4	0.00	Q_3 to Q_4	2.62
Q_4 to Q_1	0.00	Q_4 to Q_1	0.00	Q_4 to Q_1	0.00	Q_4 to Q_1	0.32
Q_4 to Q_2	0.50	Q_4 to Q_2	0.77	Q_4 to Q_2	0.00	Q_4 to Q_2	0.92
Q_4 to Q_3	1.37	Q_4 to Q_3	1.15	Q_4 to Q_3	0.00	Q_4 to Q_3	1.09

Rule	C_3 to C_4	Rule	C_4 to C_1	Rule	C_4 to C_2	Rule	C_4 to C_3
Q_1 to Q_3	0.00	Q_1 to Q_3	0.00	Q_1 to Q_3	0.00	Q_1 to Q_3	0.00
Q_1 to Q_4	0.00	Q_1 to Q_4	0.00	Q_1 to Q_4	0.00	Q_1 to Q_4	0.00
Q_2 to Q_3	0.50	Q_2 to Q_3	0.00	Q_2 to Q_3	0.00	Q_2 to Q_3	1.65
Q_2 to Q_4	1.91	Q_2 to Q_4	0.00	Q_2 to Q_4	0.00	Q_2 to Q_4	2.29
Q_3 to Q_1	0.00	Q_3 to Q_1	0.00	Q_3 to Q_1	0.00	Q_3 to Q_1	0.00
Q_3 to Q_2	0.60	Q_3 to Q_2	0.00	Q_3 to Q_2	0.00	Q_3 to Q_2	0.86
Q_3 to Q_4	3.48	Q_3 to Q_4	0.00	Q_3 to Q_4	0.00	Q_3 to Q_4	3.30
Q_4 to Q_1	0.00	Q_4 to Q_1	0.00	Q_4 to Q_1	0.00	Q_4 to Q_1	0.00
Q_4 to Q_2	0.96	Q_4 to Q_2	0.00	Q_4 to Q_2	0.00	Q_4 to Q_2	0.50
Q_4 to Q_3	1.45	Q_4 to Q_3	0.00	Q_4 to Q_3	0.00	Q_4 to Q_3	1.37

Table 7. Contribution of each association rule "$Q_i \to Q_j$" to the relevance degree of "$C_x \to C_y$" (correct-to-correct relationship)

Rule	C_1 to C_2	Rule	C_1 to C_3	Rule	C_1 to C_4	Rule	C_2 to C_1
Q_1 to Q_2	2.29	Q_1 to Q_2	0.00	Q_1 to Q_2	0.00	Q_1 to Q_2	1.59
Q_1 to Q_3	1.82	Q_1 to Q_3	0.00	Q_1 to Q_3	0.00	Q_1 to Q_3	0.00
Q_2 to Q_1	1.93	Q_2 to Q_1	0.00	Q_2 to Q_1	0.00	Q_2 to Q_1	3.07
Q_2 to Q_3	1.09	Q_2 to Q_3	0.76	Q_2 to Q_3	0.42	Q_2 to Q_3	0.00
Q_3 to Q_1	0.00	Q_3 to Q_1	0.00	Q_3 to Q_1	0.00	Q_3 to Q_1	0.00
Q_3 to Q_2	0.00	Q_3 to Q_2	0.00	Q_3 to Q_2	0.00	Q_3 to Q_2	0.00

Rule	C_2 to C_3	Rule	C_2 to C_4	Rule	C_3 to C_1	Rule	C_3 to C_2
Q_1 to Q_2	0.00	Q_1 to Q_2	0.00	Q_1 to Q_2	0.00	Q_1 to Q_2	0.00
Q_1 to Q_3	0.00	Q_1 to Q_3	0.00	Q_1 to Q_3	0.00	Q_1 to Q_3	0.00
Q_2 to Q_1	0.00	Q_2 to Q_1	0.00	Q_2 to Q_1	1.69	Q_2 to Q_1	1.53
Q_2 to Q_3	0.76	Q_2 to Q_3	0.60	Q_2 to Q_3	0.00	Q_2 to Q_3	0.86
Q_3 to Q_1	0.00	Q_3 to Q_1	0.00	Q_3 to Q_1	0.00	Q_3 to Q_1	3.30
Q_3 to Q_2	1.31	Q_3 to Q_2	0.72	Q_3 to Q_2	0.00	Q_3 to Q_2	2.38

Rule	C_3 to C_4	Rule	C_4 to C_1	Rule	C_4 to C_2	Rule	C_4 to C_3
Q_1 to Q_2	0.00	Q_1 to Q_2	0.00	Q_1 to Q_3	0.00	Q_1 to Q_3	0.00
Q_1 to Q_3	0.00	Q_1 to Q_3	0.00	Q_1 to Q_4	0.00	Q_1 to Q_4	0.00
Q_2 to Q_1	0.00	Q_2 to Q_1	1.69	Q_2 to Q_3	1.53	Q_2 to Q_3	0.00
Q_2 to Q_3	0.33	Q_2 to Q_3	0.00	Q_2 to Q_4	0.86	Q_2 to Q_4	0.60
Q_3 to Q_1	0.00	Q_3 to Q_1	0.00	Q_3 to Q_1	1.82	Q_3 to Q_1	0.00
Q_3 to Q_2	0.91	Q_3 to Q_2	0.00	Q_3 to Q_2	1.31	Q_3 to Q_2	0.91

Table 8. Final concept relevance degrees of "$C_x \to C_y$"

C_x to C_y	Degree (Path 1)	Degree (Path 2)	Final Degree	C_x to C_y	Degree (Path 1)	Degree (Path 2)	Final Degree
C_1 to C_2	2.73	2.56	2.65	C_3 to C_1	0.00	0.76	0.38
C_1 to C_3	1.40	1.64	1.52	C_3 to C_2	1.65	1.11	1.38
C_1 to C_4	2.04	1.69	1.87	C_3 to C_4	2.18	0.79	1.49
C_2 to C_1	0.00	1.89	0.95	C_4 to C_1	0.00	0.42	0.21
C_2 to C_3	2.18	2.43	2.31	C_4 to C_2	1.37	0.67	1.02
C_2 to C_4	2.38	1.47	1.93	C_4 to C_3	1.83	0.75	1.29

$$R^w_{c_1 c_2} = \sum_{r=1}^{N_w} \left(d_r^w \cdot d_r^w \right) / \sum_{r=1}^{N_w} d_r^w = \frac{3.63^2 + 2.52^2 \times 1.65^2 + 2.29^2}{3.63 + 2.52 + 1.65 + 2.29} \approx 2.73 \qquad (13a)$$

$$R^c_{c_1c_2} = \sum_{r=1}^{N_c} (d^c_r \cdot d^c_r) / \sum_{r=1}^{N_c} d^c_r = \frac{2.29^2 + 1.82^2 + 1.93^2 + 1.09^2}{2.29 + 1.82 + 1.93 + 1.09} \approx 2.56 \qquad (13b)$$

Step 5: Generate the resultant concept relevance degrees by merging results from the two paths. Here, we choose $\alpha = 0.5$ in Eq. (7). The final relevance degrees are shown in the 4th and 8th columns in Table 8, and concept maps for the concept relevance degrees in Table 8 are shown in Figs. 3, 4 and 5.

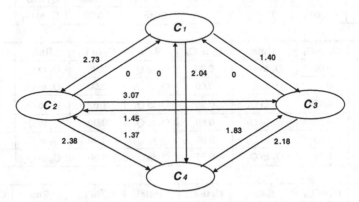

Fig. 3. A concept map generated based on the wrong-to-wrong relationship

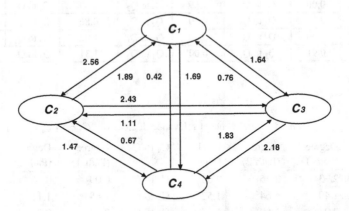

Fig. 4. A concept map generated based on the correct-to-correct relationship

3.3 Evaluation and Discussions

It is worthwhile to mention again that a concept map reveals pre-acquisition relationships between concepts. As mentioned in Sect. 3.1, the difficulty level of concepts increases from C_1 to C_4 meaning that students need to study C_1 before studying C_2, and so on. In Fig. 5, as an example, the relevance degree from C_1 to C_2 is 2.65 while that from C_2 to C_1 is 0.95. This means that C_1 is a pre-acquisition concept of C_2, and if a

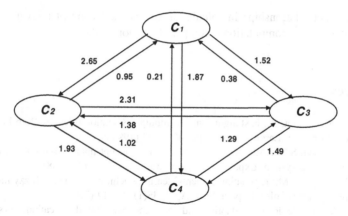

Fig. 5. A resultant concept map generated by considering both wrong-to-wrong and correct-to-correct relationships ($\alpha = 0.5$)

student fails in C_1, it is very probably that he will also fail in C_2. Similar relationships between two concepts can be found in the concept map shown in Fig. 5. That is, this proves that the concept map reflects the actual relationship among the concepts we selected.

It is also found that the concept map in Fig. 5 constructed by merging both paths can reflect the concept relevance degree more accurately than that shown in Fig. 3 which only considered the wrong-to-wrong relationship of questions. For example, in Fig. 3, the relevance degree of C_1 to C_2 is 2.73, but C_2 to C_1 is 0, which may not make sense in practical. On the other hand, in Fig. 5 that considered the correct-to-correct relationship of questions as well, the concept relevance degree of C_1 to C_2 becomes 2.65, and C_2 to C_1 becomes 0.95. Therefore, the concept map in Fig. 5 is expected to be more robust in practical.

4   Conclusions and Future Works

In this paper, we proposed a new concept map generation approach that provides more accurate relevance degrees among concepts by considering both wrong-to-wrong and correct-to-correct relationships of questions, and also by adopting accurate formulas in calculation of relevance degrees between any two concepts. We showed through simulations with practical learning courseware and test records that the proposed approach can generate a more accurate and robust concept map to improve students' learning experiences.

While it showed an enhanced accuracy for the practical data set we used, the proposed algorithm still used a small set of concepts and test records in simulations and may not work very well for test records of a very large size. Therefore, our future work will focus on improvement of the Apriori algorithm-based approach to make it work well even in actual learning systems that involve dynamic learning data of a very large size, by (i) employing more accurate function models for student test records, (ii) developing an algorithm that adaptively sets weights between wrong-to-wrong and correct-to-correct relationships in calculation of resultant concept relevance degrees, and (iii) reducing the computational complexity, among others.

References

1. Agrawal, R., Srikant, R.: Fast algorithm for mining association rules. In: 20th International Conference on Very Large Database, pp. 487–499 (1994)
2. Bai, S.M., Chen, S.M.: Automatically constructing concept maps based on fuzzy rules for adaptive learning system. Experts Syst. Appl. **35**(3), 1408–1414 (2008)
3. Bai, S.M., Chen, S.M.: Evaluating students' learning achievement using fuzzy membership functions and fuzzy rules. Experts Syst. Appl. **35**(1), 41–49 (2008)
4. Brusilovsky, P., Peylo, C.: Adaptive and intelligent web-based educational systems. Int. J. Artific. Intell. Educ. **13**, 159–172 (2003)
5. Carchiolo, V.L., Malgeri, M.: Adaptive formative paths in a web-based learning environment. Educ. Technol. Soc. **5**(4), 64–75 (2002)
6. Chen, S.M., Bai, S.M.: Using data mining techniques to automatically construct concept maps for adaptive learning systems. Expert Syst. Appl. **37**, 4496–4503 (2010)
7. Huang, X., Yang, K., Lawrence, V.B.: Classification-based approach to concept map generation in adaptive learning. In: 15th IEEE International Conference on Advanced Learning Technologies, Hualien, Taiwan (2015)
8. Lee, C.H., Lee, G.G., Leu, Y.H.: Application of automatically constructed concept map of learning to conceptual diagnosis of e-learning. Expert Syst. Appl. **36**(2), 1675–1684 (2009)
9. Liao, S.H., Chu, P.H., Hsiao, P.Y.: Data mining techniques and applications – A decade review from 2000 to 2011. Expert Syst. Appl. **39**, 11303–11311 (2012)
10. Millcevic, A.K., Vesin, B., Ivanovic, M., Budimac, Z.: E-learning personalization based on hybrid recommendation strategy and learning style identification. Comput. Educ. **56**, 885–899 (2011)
11. Novak, J.D.: Learning, creating, and using knowledge, concept maps as facilitative tools in schools and corporations. Lawrence Erlbaum and Associates, New Jersey (1998)
12. Rahman, M.A., Islam, M.Z.: A hybrid clustering technique combining a novel genetic algorithm with k-means. Knowl.-Based Syst. **71**, 345–365 (2014)
13. Sarem, M.A., Bellafkih, M., Ramdeni, M.: An approach for mining concepts' relationships based on historical assessment records. Adv. Control Eng. Inf. Sci. **15**, 3245–3249 (2011)
14. Sowan, B., Dahal, K., Hossain, M.A., Zhang, L., Spencer, L.: Fuzzy joint points based clustering algorithms for large data sets. Expert Syst. Appl. **40**, 6928–6937 (2013)
15. Tsai, C.-J., Tseng, S.S., Lin, C.-Y.: A two-phase fuzzy mining and learning algorithm for adaptive learning environment. In: Alexandrov, V.N., Dongarra, J., Juliano, B.A., Renner, R.S., Tan, C. (eds.) ICCS-ComputSci 2001. LNCS, vol. 2074, pp. 429–438. Springer, Heidelberg (2001)
16. Tseng, S., Sue, P., Su, J., Weng, J., Tsai, W.: A new approach for constructing the concept map. Comput. Educ. **49**, 691–707 (2007)
17. Wu, X., Kumar, V.: The Top Ten Algorithms in Data Mining. Chapman and Hall Publisher, Boca Raton (2009)
18. Yang, J., Huang, Z.X., Gao, Y.X., Liu, H.T.: Dynamic learning style prediction method based on a pattern recognition technique. IEEE Trans. Learn. Technol. **7**(2), 165–177 (2014)
19. Zaki, M.J., Wagner Jr., M.: Data mining and analysis: fundamental concepts and algorithms. Cambridge University Press, United Kingdom (2014)

Social Media Mining

Quantifying the Hidden Factors Impacting the Audience of Advertisements Posted on Facebook

Mamadou Diaby$^{(\boxtimes)}$ and Emmanuel Viennet

L2TI, Université Paris 13, Sorbonne Paris Cité, 93430 Villetaneuse, France
{mamadou.diaby,emmanuel.viennet}@univ-paris13.fr
http://www-l2ti.univ-paris13.fr

Abstract. The incredible current popularity of Facebook is motivating organizations to more and more use this platform to promote their products and job offers; this leads us to develop a system that estimates the audience of advertisements posted on social networks. This paper presents our proposed system that has been developed using machine learning and collaborative filtering techniques together with the data of advertisements posted on Facebook by clients of the company Work4. Our results suggest that the profile of posters, the number of persons who can see job ads, the reputation of companies, the hours, days and months of posts might impact the audience of job ads on Facebook. This work has been done at Work4, a software company that offers Facebook recruitment solutions.

Keywords: Audience of job ads · Facebook · Lasso · Regression · SVM

1 Motivations

Facebook (launched in 2004) is currently one of the most popular social networks with more than a billion of users around the world [11]. This popularity is offering organizations a great opportunity to advertise their job offers to thousands of users at the same time: this may help them to reduce their recruitment costs by speeding up their hiring process.

In this paper, we present Work4Oracle, a decision support system that predicts an estimation of the number of clicks a given advertisement should obtain; it helps recruiters optimizing the process of advertising their job offers to Facebook users by finding the right moments to post on Facebook, the right persons to post a job, the right jobs to post for a specific poster, etc. Our system has been designed by combining heterogeneous data from different sources about jobs to be advertised, their organizations and countries. The descriptions of jobs (stored

---

M. Diaby—is also a Data Scientist at the R&D Department of Work4 (http://www.work4labs.com), 75009, Paris, France.

© Springer International Publishing Switzerland 2015
P. Perner (Ed.): ICDM 2015, LNAI 9165, pp. 263–277, 2015.
DOI: 10.1007/978-3-319-20910-4_19

in the databases of Work4) generally contain information about the related positions, organizations' and countries' names, language of jobs, type of contracts (full time for instance) and requirements for the positions. We show how external sources of information, like Wikipedia or specialized websites can be used to enrich the profiles of ads; in our application, adding information (from external sources) about organizations and countries proved to be effective for the audience prediction. The analysis the results of Work4Oracle allows us to quantify factors impacting the audience of job ads posted on Facebook; we compare our findings to some studies in the literature of Human Resource Management.

The main contribution of this paper is that we quantify the hidden factors impacting the popularity of advertisements posted on Facebook using our vector model together with machine learning and collaborative filtering techniques: this allows to explain why some ads perform better than others.

We summarize work done on the factors impacting attractiveness of organizations to applicants, recommender systems and machine learning in the Sect. 2; the Sect. 3 presents the proposed methods while the Sect. 4 describes our dataset and analyses the obtained results; we end this paper by a discussion about our findings in Sect. 5.

2   Related Work

The study done in this paper is mainly related to recommender systems, machine learning and attractiveness of organizations.

2.1   Recommender Systems

Recommender systems [1,3,12] are mainly related to information retrieval, machine learning, data mining and other research fields beyond the scope of this study; they are often defined as software that elicit the interests or preferences of individual consumers for products, either explicitly or implicitly, and make recommendations accordingly [28]. Among different categories of recommender systems [1,3,13,15], we have *collaborative filtering techniques* in which recommendations for a user are based on the opinions of a community of similar users; our baseline method to estimate the audience of job ads is inspired from collaborative filtering techniques.

2.2   Machine Learning Methods

Our proposed models are mainly based on machine learning algorithms, especially regression algorithms. One of the simplest methods of regression is the ordinary least squares (OLS) in which one learns a model to minimize the sum of squares between the observed values of target variables and predicted ones; in this method, regression models in high dimensional spaces are generally subject to overfitting. In order to get a robust solution, with good generalization properties, one has to control the complexity of the model, which can be done by introducing a penalty term in the objective function [6]. Based on this observation, one of our proposed models uses Lasso regression method [24] which is especially interesting because it uses ℓ_1 prior as regularizer to reinforce the optimization towards sparse solutions (fewer non zeros parameters) which are both robust and computationally efficient. It worth noting that Lasso regression method allows to select relevant features (the weights of which are non zeros in learnt models); selecting relevant features for the prediction of the audience advertisements allows us to find out which attributes are important and to quantify their importance and then, to be able to explain to our customers why some of their ads perform better than others. We also propose two other models that are based on an adaption of Support Vector Machines [4, 6] for regression problems (ϵ-SVR); the main advantage of SVMs is their ability to build a robust and flexible non-linear models by using parametrized kernels. We use the standard form of ϵ-SVR [4, 26] with Polynomial (Poly) and RBF kernels in our experiments.

2.3 Factors Impacting Attractiveness of Organizations to Applicants

Many studies, mainly from Human Resource Management literature, have been dedicated to the identification of the factors impacting the attractiveness of organizations for people seeking jobs; we adapt the previous work in the special case of the new social media-based recruitment using machine learning methods on real-world data collected by Work4. The reference [23] investigated some factors impacting the performance of ads campaigns in recruitment context and highlighted the following important factors: *message and target of ads, type of jobs, reputation, size* and *image as employer* of the organizations and *organizations' recruitment websites*. The location, salary and description of jobs (including the industry, category and type of contract) might impact the attractiveness of job ads for future applicants [14, 16]. References [5, 10, 25] show that the reputation, size and image as employer of the organizations that propose jobs can impact the attractiveness of job ads for applicants. Recently, [27] studied how to detect the moment of a career switch for a specific user in order to recommend relevant jobs to him at that moment using hierarchical proportional hazards model together with data collected by LinkedIn.

2.4 Performance Metrics

Several performance metrics [18] have been developed in the literature to assess the performance of predictive systems: Accuracy, Precision, Recall, F_1, RMSE (Root Mean Square Error) and MAE (Mean Absolute Error). The Precision refers to the capacity of a predictive system to be precise (in the prediction of different classes) while the Recall refers to its ability to find all elements of a specific class. A predictive system can have a high Recall with a low Precision or vice versa, that's why F_β-measure [21] has been designed to take into account the Recall and the Precision; F_1 is the most often mentioned in the literature. MAE and RMSE metrics are generally used to assess the performance of regression methods. We use all these metrics in this paper to better assess the performance of our systems in both industrial and academic contexts.

3   Proposed Models

We call *audience* of a job advertisement, the number of clicks of social network users on this post. There are two ways to post job ads on Facebook using the applications developed by Work4: posts on the walls of Facebook users (Facebook-profile) and posts on organizations' Facebook pages (Facebook-page); this leads us to define our set of networks as $N = \{Facebook - profile, Facebook - page\}$. The problem we tackle in this paper can be formally stated as follows:

$$\forall n \in N, \ \Gamma_n : P \times J \times D \to \mathbb{N} \tag{1}$$

where Γ_n is an audience function for the network n, P, J are respectively the set of posters (users) and jobs, D is the set of dates and $\mathbb{N}$ is the set of all natural numbers. $\Gamma_n(p, j, d)$ represents the audience of the job j posted on the social network n by the poster p at the date d.

3.1   Modeling of Job Advertisements

Each job ad is defined by a 5-tuple (poster, job, network, date_of_post, number_of_clicks), our goal is to predict the variable *number_of_clicks* for each social network. For each job ad, we extract a profile (mainly based on binary weighting function [9]) using the information about the poster, job, network on the which the job ad has been posted and the date of the post. The extracted profile will characterize the job ad and will be used to predict its audience. Formally, we define the profile of a job advertisement as a set of:

1. the profile of its poster,
2. the profile of the associated job offer,
3. the profile of the date at which it has been or will be posted,
4. the matching vector between the poster's and job's O*NET vectors.

Profile of Posters. Each Facebook user is related to some occupations and has an associated Facebook account. We call reach of an account the number of persons who can see its posts, which corresponds to:

- the friend count of a Facebook user, if the post is done on his Facebook wall,
- the number of persons who liked the organization's Facebook page on which the job post has been done.

Due to privacy concerns, we cannot generally access the data of users connected to a specific social network user, that's why we only use the reach of accounts in this study. We could not profile more finely the users to whom the jobs are advertised but based on the principle of homophily in social networks ("Birds of a feather flock together") [17], the profile a poster (user) can give a clue on the profiles of his social connections (generally friends).

We define the profile of a user (poster) as the set of his *O*NET vector* (see Sect. 3.1) and the *vector of the reach* of his associated account.

We propose the following encoding for the reach of job ads:

$$\forall i \in \{0, ..., d-1\}\ v_i(r,d) = \begin{cases} 1 & \text{if r} > 0 \text{ and i} = \lfloor \log_{10}(r) \rfloor \\ 0 & \text{otherwise} \end{cases} \qquad (2)$$

where $v(r,d) = (v_0(r,d), ..., v_{d-1}(r,d))$ is the vector associated to the reach r (≥ 0) for a number of dimensions d. After analyzing the distribution of reaches in our dataset, we noticed that the reach values are ranging from 10^0 to 10^6 (exclusive), so we set the number of dimensions d of the vectors of reach to 6. To understand how this encoding works, let us see the following examples: 9 is encoded as $[1,0,0,0,0,0]$, 10 as $[0,1,0,0,0,0]$ and $999,999$ as $[0,0,0,0,0,1]$.

Profile of Jobs. Related work (see Sect. 2.3) showed that the attractiveness of a job generally depends on some factors like its organization's name and reputation, the salary, title and industry of the job. We extract the profile a job based on two types of data: *information about the job and its organization stored in the databases of Work4* and *additional information about the organizations and countries available on Internet.*

Modeling of Data about Jobs in the Databases of Work4: For information about jobs and their organizations stored in the databases of Work4, we define 4 sub-vectors for a specific job: the *O*NET vector* (see Sect. 3.1) linked to its title, *contract type*, *country* and *company name*. Our 4 types of contracts (Full Time, Temporary, Internship and Part Time) are encoded on 4 dimensions using binary vector model. Example: vector("Full Time") = {("Full Time", 1)}.

Similarly, our 500 distinct organizations' names and 200 targeted countries are respectively encoded on 500 and 200 dimensions using the binary weighing function.

Modeling of Data about Jobs Retrieved from Internet: We retrieve additional information about organizations and countries available on Internet from some websites (mainly on Wikipedia.org[1] and on the Organization for Economic Co-operation and Development web site (OECD)[2]) and encode them in a vector with *157 dimensions*. The *job country unemployment rate*[3] at the period (year-month) of a job post: this represents 1 dimension and we normalize its values by dividing by the max unemployment rate in our dataset; this ensures us to have values between 0 and 1. The *age of a job organization* (based on the organization creation date on Wikipedia.org) is encoded by 5 dimensions (quintiles) using binary weighting function. On organizations' Wikipedia pages, we can find their *area-served*, *types* and *industries*; our 20 distinct area-served, 20 types and 80 industries are respectively encoded on 20, 20 and 80 dimensions using binary

---

[1] http://en.wikipedia.org.

[2] http://www.oecd.org.

[3] The unemployment rates used in this paper have been retrieved from the OECD web site (only concerning OECD countries but almost all our job ads have been done for OECD countries).

weighting function. Wikipedia also generally gives the *number of employees* of an organization and its financial information (*revenue, income, operating income, asset* and *equity*). We encode the *number of employees* as the reach of posts on 6 dimensions (see Sect. 3.1); each of the financial information is encoded by 5 dimensions (quintiles) using binary weighting function.

Profile of Dates of Job Ads. The date of job posts have 6 components: year, month, day, hour, minute, second; in our study, we ignored the minute and second components of posts since they might not impact the audience of job posts. One notes that we have at most 31 days in a month but we decided to use the name of days instead since we are interested in finding the impact the day in a week on audience of job ads. We then encode a given date as a set of the vectors of its year, month, day and hour using binary vector model. For instance, knowing that *2014-07-21* corresponds to Monday and 11pm to 23 h, vector(2014-07-21:11pm) = $\{(2014, 1), (07, 1), (Monday, 1), (23, 1)\}$.

O*NET Vector of Documents. Our previous work [9] showed that the most important fields in the task of job recommendation are: *Work* for Facebook users and *Title* for jobs, as a result, we use the information contained in *Work* and *Title* fields to extract a new type of vector for Facebook users and jobs that we called O*NET vector using the O*NET-SOC taxonomy[4] [20] (a taxonomy that defines the set of occupations across the world of work). We indexed O*NET databases using Elasticsearch[5] and we thus use this library to query the different O*NET occupations (with their normalized relevance scores) related to a document (Work field data for Facebook users and title data for jobs). Each document is then represented by its distribution scores over all the occupations in O*NET databases. O*NET taxonomy currently contains 1,040 distinct occupations, as a result, our O*NET vectors are encoded by 1,040 dimensions. References [7,8] give further details about the extraction of O*NET vectors.

After extracting the O*NET vectors of posters (Facebook users) and jobs, we extract **matching vectors between posters and jobs O*NET vectors** as follows:

$$\forall i \in \{0, ..., d-1\}\ v_i(v^u, v^j) = \min(v_i^u, v_i^j) \tag{3}$$

where $v(v^u, v^j) = (v_0(v^u, v^j), ..., v_{d-1}(v^u, v^j))$ is the matching vector between v^u and v^j, $d = 1,040$ is the number of distinct O*NET occupations and v^u and v^j are respectively the O*NET vectors of a user u and job j.

3.2   Work4Oracle: A System Estimating the Audience of Job Ads

After modeling job ads and extracting their vectors, we propose our baseline method inspired from collaborative filtering recommender systems (see Sect. 2.1) called CF-Work4Oracle in which the audience of a job ad is computed as a

---

[4] http://www.onetcenter.org/taxonomy.html.

[5] http://www.elasticsearch.org.

weighted sum of the audience of similar job ads posted on the same year, month and hour on the same social network as the active job ad. Formally the audience of job ads are computed as follows:

$$\Gamma_n(\rho) = \begin{cases} 0 & \text{if } N_\rho^n \text{ is empty} \\ \dfrac{\sum_{\rho' \in N_\rho^n} \cos(\rho, \rho') \times \Gamma_n(\rho')}{\sum_{\rho' \in N_\rho^n} \cos(\rho, \rho')} & \text{otherwise} \end{cases} \tag{4}$$

where ρ and ρ' are vectors of job ads, $\Gamma_n(\bullet)$ is the audience of $\bullet$, N_ρ^n is a set of all job posts similar (similarity > 0) to ρ posted on the network n at the same year, month and hour as ρ and cos is the cosine similarity [9].

After defining the baseline method, we design and study 3 systems based on regression methods (see Sect. 2.2): sPoly-Work4Oracle, sRBF-Work4Oracle and Lasso-Work4Oracle which respectively use ϵ-SVR with Polynomial kernel, ϵ-SVR with RBF kernel and Lasso regression to learn models to predict the audience of job ads posted on Facebook.

The numbers of clicks on job ads in our dataset are ranging from 0 to 2086. Our preliminary experiments have showed that it is to difficult to fit a model to accurately predict target values between 0 and 2086. We decide to use the log-scaled number of clicks (log(1 + number of clicks) where log is the natural logarithm) to fit our models; in that case target values are ranging from 0 to 7.64. Since we use log-scaled number of clicks to fit our models, the predictions (integer) are obtained as follows:

$$\hat{y}(x) = \max(0, \lfloor -1 + \exp^{\text{model}(x)} \rceil) \tag{5}$$

where $\lfloor \bullet \rceil$ is the nearest integer to $\bullet$ and *model* is a model learnt from our data using log-scaled number of clicks.

4 Experiments

This section presents our dataset about the performance of job ads on Facebook, the experiments we have conducted and the obtained results.

4.1 Description of Our Dataset

We evaluate the performance of our proposed systems on a dataset collected by Work4 between February 2013 and June 2014. Each entry in the dataset is a 5-tuple (u, j, n, d, y) where u and v are a social network user (poster) and job respectively and $n \in \{Facebook - profile, Facebook - page\}$ is the network on which the job has been posted; d is the date of the post and y is the number of clicks on the post. We clean up our dataset by removing posts from the Work4's developers and testers; finally we obtain a dataset with 33,366 job ads. Table 1 reports the summary statistics of our dataset. For a reminder, each job ad is associated to a social network account that has a specific reach (number

Table 1. Statistics extracted from our dataset; the numbers of posts, posters and job; we have computed the quintiles of the number of clicks obtained by our job advertisements. Facebook-page and Facebook-profile correspond to posts on organizations' Facebook pages and Facebook users' walls respectively.

	Facebook-page	Facebook-profile	Total
Number of users	512	3,376	-
Number of jobs	14,691	9,030	-
Number of job pages	490	1,084	-
Number of posts in 2013	7,810	4,418	**12,228**
Number of posts in 2014	12,583	8,555	**21,138**
Total number of posts	20,393	12,973	**33,366**
1st quintile	1	1	
2nd quintile	2	1	Number of clicks
3rd quintile	4	1	
4th quintile	9	2	

of persons who can see the posts of this account). Statistics from our dataset reveal that the number of persons who can see the job ads posted on organizations' Facebook pages (7,277 on average) is by far higher than the number of persons who can see the job posted on the walls of Facebook users (542 on average). We call half-life of job ads, the number of hours (after its post on a social network) required to get the half of the total number of clicks it obtained. Figure 1 compares the half-life of job advertisements posted on Facebook users' walls and organizations' Facebook pages; it reveals that the half-life of job posts on Facebook users' walls is very short (about 4 h) compared to the half-life of those posted on organizations' Facebook pages (more than 96 h).

4.2 Results

Our scripts are written in Python and are mainly based on scikit-learn[6] [19] implementation of different regression algorithms and performance metrics (the implementation of SVM-R is based on LIBSVM [4]). The different experiments have been run on Intel Xeon 2.00 GHz (with 12 cores). For all proposed methods (CF-Work4oracle, sPoly-Work4oracle, sRBF-Work4oracle and Lasso-Work4Oracle), we learn a model and test it for different social networks separately using 5-fold cross-validation. We precompute the similarities between job ads in order to speed up our experimentations with CF-Work4Oracle.

Recall that a linear learnt model is defined (in our context) by:

$$\left(\alpha, \ (w_i^f)_{\substack{1 \leq f \leq n_f \\ 1 \leq i \leq k_f}} \right) \tag{6}$$

---

[6] http://scikit-learn.org.

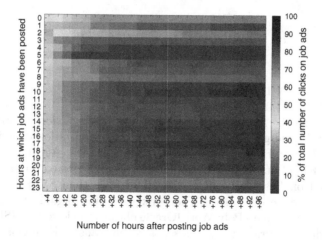

(a) Job ads posted on walls of Facebook users.

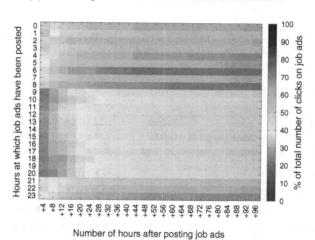

(b) Job ads posted on organizations' Facebook pages.

Fig. 1. Evolution of percentage of total number of clicks on job ads posted on Facebook. By defining the half-life of job ads as the number of hours required to get the half of the total number of clicks it obtained, we note that the half-life of job ads posted on the walls of Facebook users is by far shorter than the half-life of those posted on organizations' Facebook pages.

where α is the bias, w_i^f is the learnt weight for the component i of the feature f, n_f is the number of different features in the model and k_f is the number of dimensions associated to the feature f. Based on the equation (6), we calculate the contribution c_f of a feature f using non-zeros learnt parameters as follows:

$$c_f = \begin{cases} \dfrac{\sum_{i=1}^{k_f} |w_i^f|}{\sum_{i=1}^{k_f} 1_{w_i^f \neq 0}} & \text{if } (\sum_{i=1}^{k_f} 1_{w_i^f \neq 0}) \neq 0 \\ 0 & \text{otherwise} \end{cases} \tag{7}$$

where $1_{\text{condition}} = \begin{cases} 1 & \text{if condition is true} \\ 0 & \text{otherwise} \end{cases}$

We optimize the hyper-parameters of different regression algorithms using cross-validation and grid-search (similarly to the optimization done in [22]).

Table 2 depicts the training times, RMSE, MAE, Accuracy, Precisions, Recalls and F1s of our proposed methods for job ads posted on Facebook. The models based on Lasso take more time fit models than SVM-based methods for both posts done on walls of Facebook users and organizations' Facebook pages. We can note that sPoly-Work4Oracle obtains the lowest RMSE and MAE and the highest accuracy for job ads posted on walls of Facebook users while Lasso-Work4Oracle and sRBF-Work4Oracle have better results for job posted on organizations' Facebook pages.

Table 2. Training time, RMSE, MAE and accuracy of Work4Oracle for different regression methods for job ads on Facebook using 5-fold cross-validation.

	Methods	Training time (seconds)	RMSE	MAE	Accuracy
Job ads on Facebook users' walls	CF-Work4Oracle	-	2.47±0.12	1.14±0.01	0.49±0.01
	sPoly-Work4Oracle	31.65±0.29	**2.33±0.18**	**0.81±0.03**	**0.66±0.01**
	sRBF-Work4Oracle	31.94±0.52	2.36±0.16	1.15±0.03	0.53±0.01
	Lasso-Work4Oracle	1988.54± 7.92	2.41±0.40	0.93±0.04	0.59±0.01
Job ads on organizations' Facebook pages	CF-Work4Oracle	-	50.85±7.09	15.50±0.85	0.49± 0.01
	sPoly-Work4Oracle	63.39±1.32	39.71±6.96	8.76±0.42	0.72±0.02
	sRBF-Work4Oracle	94.70±1.46	38.91±8.55	**7.84±0.66**	**0.78±0.01**
	Lasso-Work4Oracle	2,696.29±13.52	**37.48±7.54**	7.87±0.55	**0.78±0.01**

A fine analysis of the Table 3 for job ads posted on Facebook users' walls shows that sPoly-Work4Oracle has a higher value of F1 for job ads with at most 1 click (third quintile of the number of clicks) but a very low F1 on job ads with at least 2 clicks: this model is not efficient in detecting job ads (on Facebook users' walls) with at least 2 clicks; Lasso-Work4Oracle makes a better trade-off between precision and recall. For job ads posted on organizations' Facebook pages, we obtain highest F1 scores for sRBF-Work4Oracle and Lasso-Work4Oracle. CF-Work4Oracle is globally outperformed by the others for both job ads on Facebook users' walls and on organizations' Facebook pages: using learnt models allows to improve the quality of estimation of the number of clicks of job ads on Facebook than heuristic-based methods.

Table 3. Precision, recall and F1 of Work4Oracle for different regression methods for job advertisements on Facebook.

Number of clicks	≤ 1 (1: 3rd quintile)	≥ 2
Precision CF-Work4Oracle	0.69±0.01	0.33±0.01
sPoly-Work4Oracle	0.70±0.01	0.45±0.03
sRBF-Work4Oracle	0.71±0.01	0.35±0.01
Lasso-Work4Oracle	0.73±0.01	0.39±0.02
Recall CF-Work4Oracle	0.45±0.01	0.57±0.01
sPoly-Work4Oracle	0.90±0.01	0.18±0.02
sRBF-Work4Oracle	0.51±0.02	0.57±0.02
Lasso-Work4Oracle	0.62±0.02	0.52±0.02
F1 CF-Work4Oracle	0.54±0.01	0.42±0.01
sPoly-Work4Oracle	**0.78±0.01**	0.25±0.02
sRBF-Work4Oracle	0.59±0.02	**0.44±0.01**
Lasso-Work4Oracle	0.67±0.01	**0.45±0.02**

(a) Job advertisements posted on Facebook users' walls.

Number of clicks	≤ 4 (4: 3rd quintile)	≥ 5
Precision CF-Work4Oracle	0.73±0.01	0.40±0.00
sPoly-Work4Oracle	0.78±0.02	0.62±0.03
sRBF-Work4Oracle	0.81±0.01	0.72±0.02
Lasso-Work4Oracle	0.82±0.01	0.70±0.01
Recall CF-Work4Oracle	0.33±0.01	0.79±0.00
sPoly-Work4Oracle	0.79±0.03	0.60±0.02
sRBF-Work4Oracle	0.86±0.01	0.65±0.01
Lasso-Work4Oracle	0.84±0.01	0.68±0.01
F1 CF-Work4Oracle	0.45±0.01	0.53±0.00
sPoly-Work4Oracle	0.78±0.02	0.61±0.02
sRBF-Work4Oracle	**0.84±0.01**	**0.68±0.01**
Lasso-Work4Oracle	**0.83±0.00**	0.69±0.02

(b) Job advertisements posted on organizations' Facebook pages.

Now, let us analyze the contributions of different factors. First of all, we obtain a better trade-of between precision and recall with Lasso-Work4oracle (on Facebook users' walls) and best results for job ads posted on organization Facebook pages using Lasso-Work4oracle and sRBF-Work4Oracle; we then compute the contributions of different factors (see Fig. 2) using the Eq. (7) together with the models learnt by Lasso-Work4oracle.

The analysis of the different contributions of factors reveals that the audience of job advertisements posted on the walls of Facebook users depends on different factors (see Fig. 2a), the most important of which are: *profile of posters/matching between the profiles of posters and jobs, profile of jobs* (and types of contract), *name of companies* (eventually their reputations), *name of countries* (and their unemployment rates) and *hours of job posts*.

For the job ads posted on organizations' Facebook pages (see Fig. 2b), we find out that the most important factors are: *reach of job posts, name of companies* (and eventually their reputations), *countries of jobs* (and their unemployment rates) and *profile of jobs*.

Another round of analysis of our results reveal that more the reach of a post is large, higher the number of clicks is for both posts on organizations' Facebook pages and Facebook users' walls. They also show that the months, days and hours of posts impact their audience. For instance, we find that job ads posted during weekends obtain less clicks than the others for both Facebook users' walls and organizations' Facebook pages.

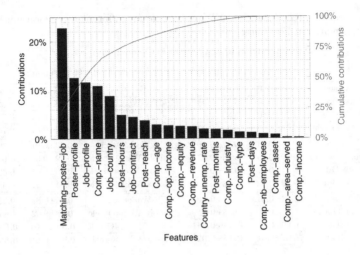

(a) Job ads posted on Facebook users' walls.

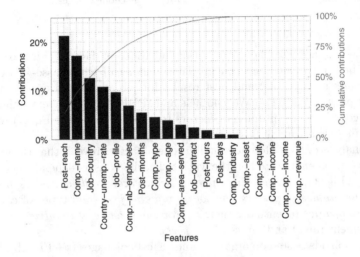

(b) Job ads posted on Organizations' Facebook pages.

Fig. 2. Percentage of contributions (as defined by the Eq. (7)) of different features to the performance of job advertisements posted on Facebook.

5   Discussion and Future Work

We have started this paper by studying the literature of human resource management: previous work found out some important factors impacting the attractiveness of organizations for applicants but these studies are based on small sets of data and their conclusions are generally based on subjective analysis which limits their generalization. We combined heterogeneous data from Work4, Wikipedia and OECD websites and defined a list of features that could be important in the task estimating the attractiveness of organizations for applicants and use them to propose a vector model for job ads based on the taxonomy O*NET and the vector space model with the binary weighting function and making sure that the values of all components are ranging from 0 to 1.

We then applied data mining and machine learning techniques to a significant large real-world dataset collected by the company Work4 to find out and quantify the factors impacting the audience of job advertisements on Facebook. Our results show that the number of clicks obtained by job advertisements posted on Facebook depend on some hidden factors, the most important are the number of persons who can see the ads, the profile of posters (which can give a clue on the profiles of friends of the posters who will see the posts), the name of the organizations of jobs (which gives also a clue on their reputations), the size of job companies, the countries (and their unemployment rates) for which the job ads have been posted, the profiles of jobs (their functions, industries and categories) and the months, hours and days at which job ads have been posted. These results confirm the findings of [5, 10, 14, 16, 25] who support that location, salary and description of jobs, the reputation, size and image as employer of the organizations may impact the attractiveness of job ads for applicants: our results extend these previous work to the social media-based recruitments.

The number of persons who can see the job ads (reach of posts) is the most important factor that impacts the audience of job ads on organizations' Facebook pages (see Fig. 2). Only persons who liked an organization page can see its job ads, so we can conclude that the popularity of an organization on Facebook affects the audience of its job ads. The notion of popularity of an organization may be closely linked to its reputation: an organization with a good reputation can be popular for future applicants. We find out that the months of posts generally impact more the performance of job ads than the hours of posts which have a higher impact than the days at which job ads have been posted (see Fig. 2). Our experiments showed that the hour of posts impact more for job ads posted on the walls of Facebook users than for jobs posted on organizations' Facebook pages; this could be explained by the fact that the half-life of job posts on the walls of Facebook users is very short (see Fig. 1).

We have recently collected two datasets similar to the one used in this paper: one about job ads posted on LinkedIn and the other about job ads posted on Twitter. In our future work we will conduct experiments similar to those we conducted in paper to find out and quantify the factors impacting the audience of advertisements posted on LinkedIn and Twitter and other social networks. We'll also be working to improve the quality of predictions of Work4Oracle by investigating the use of other machine learning algorithms like deep learning [2].

Acknowledgments. This work is supported by Work4, ANRT (ANRT: Association Nationale de la Recherche et de la Technologie.) (the French National Research and Technology Association) under the grant N°2012/0365 and French FUI Project AMMICO and the project Open Food System. (ANRT: Association Nationale de la Recherche et de la Technologie.)

References

1. Adomavicius, G., Tuzhilin, A.: Toward the next generation of recommender systems: a survey of the state-of-the-art and possible extensions. IEEE Trans. Knowl. Data Eng. **17**(6), 734–749 (2005). ISSN 1041–4347
2. Bengio, Y.: Learning deep architectures for ai. Found. Trends Mach. Learn. **2**(1), 1–127 (2009). ISSN 1935–8237
3. Bobadilla, J., Ortega, F., Hernando, A., Gutiérrez, A.: Recommender systems survey. Knowl. Based Syst. **46**, 109–132 (2013)
4. Chang, C.-C., Lin, C.-J.: Libsvm: a library for support vector machines. ACM Trans. Intell. Syst. Technol. **2**(3), 27:1–27:27 (2011). ISSN 2157–6904. Software available at http://www.csie.ntu.edu.tw/cjlin/libsvm
5. Chapman, D.S., Uggerslev, K.L., Carroll, S.A., Piasentin, K.A., Jones, D.A.: Applicant attraction to organizations and job choice: a meta-analytic review of the correlates of recruiting outcomes. J. Appl. Psychol. **90**(5), 928 (2005)
6. Cortes, C., Vapnik, V.: Support-vector networks. Mach. Learn. **20**(3), 273–297 (1995). ISSN 0885–6125
7. Diaby, M., Viennet, E.: Taxonomy-based job recommender systems on facebook and linkedin. In: Proceedings of the 2014 IEEE Eighth International Conference on Research Challenges in Information Science RCIS 2014, pp. 237–244. IEEE, May 2014
8. Diaby, M., Viennet, E.: Job recommendations on social networks using a multilayer vector model. In: Workshop on Heterogeneous Information Access at WSDM 2015 (HIA 2015), Shanghai, February 2015
9. Diaby, M., Viennet, E., Launay, T.: Exploration of methodologies to improve job recommender systems on social networks. Soc. Netw. Anal. Min. **4**(1), 227 (2014). doi:10.1007/s13278-014-0227-z. ISSN 1869–5450. http://dx.doi.org/10.1007/s13278-014-0227-z
10. Ehrhart, K.H., Ziegert, J.C.: Why are individuals attracted to organizations? J. Manag. **31**(6), 901–919 (2005)
11. Facebook, April 2015. http://newsroom.fb.com/company-info/
12. Jannach, D., Zanker, M., Felfernig, A., Friedrich, G.: Recommender Systems: An Introduction. Cambridge University Press, New York (2010)
13. Kazienko, P., Musial, K., Kajdanowicz, T.: Multidimensional social network in the social recommender system. IEEE Trans. Syst. Man Cybern. Part A: Syst. Hum. **41**(4), 746–759 (2011)
14. Lievens, F., Highhouse, S.: The relation of instrumental and symbolic attributes to a company's attractiveness as an employer. Pers. Psychol. **56**(1), 75–102 (2003). ISSN 1744–6570
15. Lops, P., de Gemmis, M., Semeraro, G.: Content-based recommender systems: state of the art and trends. In: Ricci, F., Rokach, L., Shapira, B., Kantor, P.B. (eds.) Recommender Systems Handbook, pp. 73–105. Springer, Heidelberg (2011). ISBN 978-0-387-85819-7
16. Mathews, B.P., Redman, T.: Managerial recruitment advertisements-just how market orientated are they? Int. J. Sel. Assess. **6**(4), 240–248 (1998). ISSN 1468–2389
17. McPherson, M., Smith-Lovin, L., Cook, J.M.: Birds of a feather: homophily in social networks. Ann. Rev. Sociol. **27**(1), 415–444 (2001). doi:10.1146/annurev. soc.27.1.415
18. Omary, Z., Mtenzi, F.: Machine learning approach to identifying the dataset threshold for the performance estimators in supervised learning. Int. J. Infonomics (IJI) **3**(9), 314–325 (2010)

19. Pedregosa, F., Varoquaux, G., Gramfort, A., Michel, V., Thirion, B., Grisel, O., Blondel, M., Prettenhofer, P., Weiss, R., Dubourg, V., et al.: Scikit-learn: machine learning in python. J. Mach. Learn. Res. **12**, 2825–2830 (2011)

20. Peterson, N.G., Mumford, M.D., Borman, W.C., Richard Jeanneret, P., Fleishman, E.A., Levin, K.Y., Campion, M.A., Mayfield, M.S., Morgeson, F.P., Pearlman, K., Gowing, M.K., Lancaster, A.R., Silver, M.B., Dye, D.M.: Understanding work using the occupational information network (o* net): implications for practice and research. Pers. Psychol. **54**(2), 451–492 (2001). ISSN 1744-6570

21. Van Rijsbergen, C.J.: Information Retrieval, 2nd edn. Butterworth-Heinemann, Newton (1979). ISBN 0408709294

22. Szepannek, G., Gruhne, M., Bischl, B., Krey, S., Harczos, T., Klefenz, F., Dittmar, C., Weihs, C.: Perceptually based phoneme recognition in popular music. In: Locarek-Junge, H., Weihs, C. (eds.) Classification as a Tool for Research, pp. 751–758. Springer, Heidelberg (2010)

23. Séguela, J.: Fouille de données textuelles et systèmes de recommandation appliqués aux offres d'emploi diffusées sur le web. Ph.D. thesis, Conservatoire National des Arts et Métiers (CNAM), Paris, France, May 2012

24. Tibshirani, R.: Regression shrinkage and selection via the lasso. J. R. Stat. Soc. B **58**, 267–288 (1994)

25. Turban, D.B., Forret, M.L., Hendrickson, C.L.: Applicant attraction to firms: influences of organization reputation, job and organizational attributes, and recruiter behaviors. J. Vocat. Behav. **52**(1), 24–44 (1998). ISSN 0001-8791

26. Vladimir, N.V.: Statistical Learning Theory. Wiley-Interscience, New York (1998)

27. Wang, J., Zhang, Y., Posse, C., Bhasin, A.: Is it time for a career switch? In: Proceedings of the 22nd International Conference on World Wide Web, pp. 1377–1388. International World Wide Web Conferences Steering Committee (2013)

28. Xiao, B., Benbasat, I.: E-commerce product recommendation agents: use, characteristics, and impact. Mis Q. **31**(1), 137–209 (2007)

Author Index

Printed in the United States
By Bookmasters